Research Report

Fuel Reduction for the Mobility Air Forces

Christopher A. Mouton, James D. Powers, Daniel M. Romano, Christopher Guo, Sean Bednarz, Caolionn O'Connell

RAND Project AIR FORCE

Prepared for the United States Air Force
Approved for public release; distribution unlimited

RAND
CORPORATION

For more information on this publication, visit www.rand.org/t/RR757

Library of Congress Control Number: 2015931912

ISBN: 978-0-8330-8765-2

Published by the RAND Corporation, Santa Monica, Calif.

© Copyright 2015 RAND Corporation

RAND® is a registered trademark.

Support RAND
Make a tax-deductible charitable contribution at
www.rand.org/giving/contribute

www.rand.org

Preface

The Department of Defense (DoD) is the largest U.S. government user of energy, typically accounting for about 80 percent of the total government energy consumption. In particular, aviation fuel accounts for about half of DoD's total energy use. In the context of significant fuel price increases—DoD spending on petroleum rose 381 percent between fiscal year (FY) 2005 and FY 2011 whereas DoD's petroleum use decreased 4 percent over the same period—the current Air Force Energy Plan aims to reduce the consumption of aviation fuel 10 percent from a 2006 baseline by 2015. As of March 2013, the Air Force had already exceeded that goal, achieving a 12 percent reduction.

To help achieve this goal, energy efficiency and conservation measures have been implemented during this time, and additional initiatives are being considered as a way to further reduce fuel consumption. These measures include technology improvements (aerodynamics, aircraft weight, propulsion, etc.) and fleet, flight, and ground operations optimization. This report provides a detailed analysis of the fuel and cost savings potential of fuel efficiency initiatives being considered by Air Mobility Command (AMC), the biggest consumer of aviation fuel in DoD.

A summary of this report is also available:

Christopher A. Mouton, James D. Powers, Daniel M. Romano, Christopher Guo, Sean Bednarz, and Caolionn O'Connell, *Fuel Reduction for the Mobility Air Forces: Executive Summary*, Santa Monica, Calif.: RAND Corporation, RR-757/1-AF, 2015.

The research reported here was commissioned by the AMC Director of Operations and conducted within the Resource Management Program of RAND Project AIR FORCE as part of a project entitled "Reducing Mobility Air Forces (MAF) Energy Costs: Best Practices in Aviation Operations and Training." It should be of interest to mobility air operations planners and those concerned with energy use within DoD.

RAND Project AIR FORCE

RAND Project AIR FORCE (PAF), a division of the RAND Corporation, is the U.S. Air Force's federally funded research and development center for studies and analyses. PAF provides the Air Force with independent analyses of policy alternatives affecting the development, employment, combat readiness, and support of current and future air, space, and cyber forces. Research is conducted in four programs: Force Modernization and Employment;

Manpower, Personnel, and Training; Resource Management; and Strategy and Doctrine. The research reported here was prepared under contract FA7014-06-C-0001.

Additional information about PAF is available on our website: http://www.rand.org/paf/

Contents

vi

Figures

Tables

Summary

Reducing aviation fuel use has been an ongoing goal for military and civil operators, and there is an extensive literature on the topic. As early as 1976, the Military Airlift Command (MAC), now Air Mobility Command (AMC), published a pamphlet entitled "Birds Fly Free, MAC Doesn't."[1] Although the material is presented in a humorous way, the topics it discusses are still relevant today, and many of the fuel saving concepts presented therein are revisited in our work. This is a reminder that fuel efficiency is not necessarily about groundbreaking new ideas; rather, it is about consistently implementing and following known best practices.

AMC consumed just over half of all aviation fuel used by the Air Force in fiscal year 2008.[2] As a result, there has been increasing pressure for AMC to seek opportunities to reduce fuel use. To this end, the AMC Fuel Efficiency Office was chartered in 2008 to identify and implement opportunities for fuel reduction.[3] As part of the increased emphasis on fuel efficiency, the Air Force set a goal to reduce fuel consumption by 10 percent from a 2006 baseline by 2015.[4] In March 2013, the Air Force had already achieved a 12 percent reduction.[5] Although this goal has been met, it is still prudent for the Air Force to pursue cost-effective options to further reduce fuel use.

The literature on fuel use in the aviation industry is extensive, but there are two difficulties applying this to AMC. First, most existing literature does not broadly calculate savings at the enterprise level. Second, existing literature focuses mainly on commercial operations. One intention of this report is to determine when applying commercial standards to the Air Force is appropriate and to calculate the enterprise-level savings such practices would impart.

After reviewing academic research and existing fuel reduction initiatives in the Air Force and industry, we developed a list of fuel reduction options. Some are not likely to be cost-effective for the Air Force, but their inclusion is important to develop the total potential for fuel savings regardless of cost. This list was constructed after a review of commercial practices and aviation

---

[1] Military Airlift Command, Navigation and Performance Division of Aircrew Standardization, "Birds Fly Free, MAC Doesn't," pamphlet, Scott AFB, Ill.: Navigation and Performance Division of Aircrew Standardization, February 10, 1976.

[2] Laura McAndrews, "Fuel Efficiency Among Top Priorities in AMC's Energy Conservation," Scott Air Force Base, Ill.: Air Mobility Command Public Affairs, October 5, 2009.

[3] Scott T. Sturkol, "AMC Fuel Efficiency Office Shows How 'Efficiency Promotes Effectiveness,'" Scott Air Force Base, Ill.: Air Mobility Command Public Affairs, January 5, 2011.

[4] United States Department of Energy, *Air Force Achieves Fuel Efficiency through Industry Best Practices,* Washington, D.C., DOE/GO-102012-3725, December 2012.

[5] Jared Serbu, "Air Force Meets Fuel Efficiency Goal Several Years Early," *FederalNewsRadio.com,* March 22, 2013; and Tech. Sgt. Matthew Bates, "Every Drop Counts," Travis AFB, Calif.: Defense Media Activity, November 1, 2013.

literature, which is discussed in Chapter Two. Table S.1 shows the list of the options we considered, in the order presented in this document.

Our work aims to quantify the fuel savings potential of these options for the AMC enterprise. Accordingly, we adapt the methodologies and data found in the literature to AMC operations, making appropriate adjustments in our analysis to account for the differences between military aviation and commercial airlines. We are acutely aware that the Air Force differs in many fundamental ways from a commercial airline, and we make appropriate adjustments in our analysis to account for these differences.

Table S.1. Fuel Reduction Options Analyzed

Engine-out Taxiing	Optimum Flight Level and Speed
Basic Weight Reduction	Auxiliary Power Unit Use Reduction
Load Balancing Improvement	Technical Stop Addition
Continuous Descent	Vortex Surfing
Paint Reduction	Microvanes
Ground Towing	Lift Distribution Control
Engine Modification or Replacement	Winglets
Riblets	New Aircraft

Overall Results

Of the 16 options for reducing fuel use we considered, 12 were cost-effective, i.e., the annualized implementation cost per gallon saved is less than the current price of fuel per gallon. However, half have significant negative implications to implementation. This leaves six options that are both cost-effective and can be reasonably implemented. These options are engine-out taxiing, flying optimum flight level and speed, basic weight reduction, auxiliary power unit (APU) use reduction, load balancing improvement, and microvanes.[6]

These results are summarized in Figure S.1. The x-axis of the figure represents how much fuel the option saves, i.e., the width of each bar gives the fuel savings of that option. The y-axis is the cost to implement the savings option; therefore, the height of each option represents the cost of implementing the reduction option. As an example, ground towing of C-17 aircraft would save 13.9 million gallons (MG) of fuel a year and cost $2.18 per gallon.

The figure shows a couple of significant and important trends. First, of all the fuel reduction options, those that are cost-effective—below the dashed line—represent approximately 50 MG

---

[6] Lift distribution control is also cost-effective and can be reasonably implemented; however, if microvanes are installed, the benefit of lift distribution control will likely be significantly reduced.

Figure S.1. Cost-Effectiveness of Fuel Reduction Options

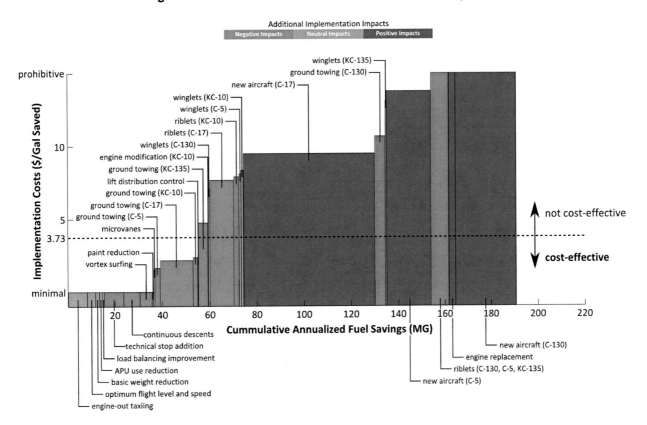

of fuel reduction a year. However, the options that are not cost-effective represent a much bigger potential for fuel reduction, 135 MG. This illustrates that although many possibilities for fuel reduction exist, only a portion of the savings potential is cost-effective. Of the options that are cost-effective, half have negative implementation impacts (orange shading). In fact, the cost-effective options that have neutral or positive implementation impacts represent only 16 MG of fuel, or 1.6 percent of projected fuel use by Mobility Air Forces. This 16 MG of fuel savings is possible while maintaining the programmed flight hours. Reductions in flight hours, for example by replacing flight hours with simulator training or more effectively programming the flight hours, would, of course, yield additional savings.

We also see in the figure a clustering of options. There are four broad categories: (1) a small set of options that are both cost-effective and have neutral implementation impacts (tan shading), (2) a set of options that are cost-effective but have negative implementation impacts, (3) a set of options that are not cost-effective and have negative implementation impacts, and (4) a set of options that are decidedly not cost-effective yet have positive implementation impacts (dark green shading).

Illustrative Examples

We present three exemplar fuel saving options that demonstrate some much broader conclusions. First, winglets are an excellent example of a technology that is cost-effective for the airlines but not for the Air Force, primarily because of the total number of hours each aircraft flies per year. Second, vortex surfing illustrates how some options are unique to the Air Force. Vortex surfing could be adopted by commercial industry, but several barriers exist for industry, such as ride quality and overlapping flight paths, but not as much for the Air Force. Finally, engine-out taxiing illustrates an option that is widely used in industry but is currently limited in Air Force use. Such an option provides significant fuel savings without significant implementation barriers.

Winglets

We find many examples of fuel efficiency initiatives that are cost-effective for the commercial airline industry but not for the Air Force. This difference arises mainly because Air Force aircraft fly significantly fewer hours per year. To understand the implications of this fact, we can consider the introduction of winglets into the commercial fleet. Figure S.2 shows the approximate price of fuel from 2000 to 2008 in FY 2013 dollars along with the approximate time that airlines began retrofitting their fleets. Many of the retrofits occurred when fuel prices were around $1.50 per gallon. If we assume that, on average, commercial carriers installed winglets when fuel prices reached the point where modernization was cost-effective for their fleet (i.e., the cost avoidance resulting from more efficient cruise exceeded the cost of adding winglets), then we can estimate the fuel price for which adding winglets becomes cost-effective for AMC. Depending on the estimate, commercial airlines may fly as many as 4,400 hours per aircraft per year.[7] According to the FY 2013 President's Budget programmed flight hours for FY 2014, aircraft in the MAF fleet are expected to fly between slightly fewer than 300 and slightly more than 800 flight hours per year, depending on the mission design series (MDS).[8] A simple calculation then tells us that we might expect winglets to be cost-effective for the MAF fleet starting at $8.25 per gallon all the way up to $22 per gallon, depending on the MDS. Our detailed analysis of winglets shows that the actual results are consistent with this approximation.

Vortex Surfing

Vortex surfing is a form of formation flying designed to reduce fuel use. Specifically, an aircraft can fly in the vortices generated by a leading aircraft. Aircraft flying in the outboard portion of the vortex would experience an upwash generated by these vortices; this would reduce

---

[7] Airbus, "The A330/A340 Family Jetliners Benefit from Lower Maintenance Costs," press release, April 16, 2009.

[8] United States Department of Defense, *Fiscal Year (FY) 2013 President's Budget: Flying Hour Program (PA)*, an extract of the Programmed Data System (PDS), Washington, D.C., February 2012, not available to the general public.

Figure S.2. Airline Winglet Retrofits over Time Based on Fuel Price

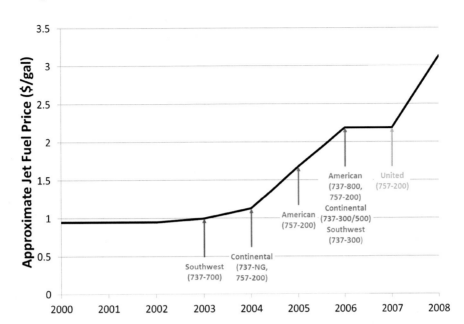

SOURCE: Adapted from Daniel M. Norton, Donald Stevens, Yool Kim, Scott Hardiman, Somi Seong, Fred Timson, John Tonkinson, Duncan Long, Nidhi Kalra, Paul Dreyer, Artur Usanov, Kay Sullivan Faith, Benjamin F. Mundell, and Katherine M. Calef, *An Assessment of the Addition of Winglets to the Air Force Tanker Fleets,* Santa Monica, Calif.: RAND Corporation, MG-895-1-AF, January 2012, not available to the general public.

the amount of thrust required to maintain level flight, thereby producing fuel savings. The configuration is shown in Figure S.3.

Figure S.3. Vortex Surfing Configuration

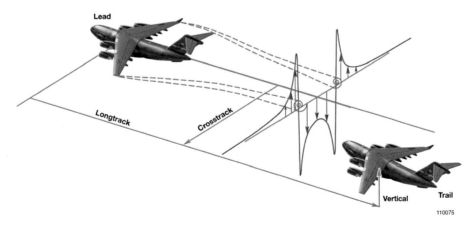

SOURCE: Joe Pahle et al., "A Preliminary Flight Investigation of Formation Flight for Drag Reduction on the C-17 Aircraft," briefing, Washington, D.C.: National Aeronautics and Space Administration, NASA 20120007201, March 7, 2012.

Flight tests conducted in 2011, 2012, and 2013 as part of two programs known as Cargo Aircraft Precision Formations for Increased Range and Efficiency (CAPFIRE) and Surfing Aircraft Vortices for Energy ($AVE) have shown cruise fuel flow reductions upward of 10 percent.[9] Initial indications are that the ride is rougher but acceptable, and damaging effects on the airframe have not been observed.[10] This rougher ride may very well be unacceptable for commercial airlines, whereas it may be tolerable for the Air Force.

To take advantage of vortex surfing, two aircraft must be flying the same route at the same time. This is much more common for the Air Force than for any single airline, again making the Air Force a better candidate for vortex surfing adoption.

Our results show that for the C-17, 15 percent of the total miles flown can be flown while trailing another aircraft. Applying an 8 percent cruise fuel flow reduction to these sorties nets a 5.3 percent total fuel burn reduction for the trailing aircraft.[11] Across the entire C-17 fleet, this translates to a reduction of 0.8 percent in fuel use.

Engine-Out Taxiing

Aircraft can often taxi using fewer than all engines and thereby reduce their fuel burn without additional investment. This occurs because even the idle thrust of large turbofan engines is sufficient to move an aircraft at taxi speed on the ground. Commercial carriers regularly use one engine when taxiing.[12] For example, JetBlue reports that it is able to conduct single-engine taxiing on 62 percent of its A320 flights systemwide.[13]

Engine-out taxiing, which places only a limited burden on maintenance and crew operations, can potentially save 8.1 MG of fuel, or $30.3 million at today's fuel price of $3.73 per gallon. Most of these savings come from the C-17 fleet and its high annual sortie rates.

Recommendations

We make several recommendations to further AMC's fuel consumption reduction efforts based on the analysis presented here. We break these recommendations into two categories. The first includes cost-effective actions that the Air Force can take now to reduce fuel use and begin to save 16 MG of fuel a year, which is about 1.6 percent of total MAF fuel consumption. The

---

[9] Pahle et al., 2012; Roger Drinnon, "'Vortex Surfing' Could Be Revolutionary," *Air Force Print News Today,* October 10, 2012; Paul D. Shinkman, "'Vortex Surfing' Could Save Military Millions," *U.S. News and World Report,* August 13, 2013; and Joe Pappalardo, "Vortex Surfing: Formation Flying Could Save the Air Force Millions on Fuel," *Popular Mechanics,* July 17, 2013.

[10] Pahle et al., 2012.

[11] We use 8 percent rather than 10 percent because there is no fuel reduction associated with start, taxi, takeoff, and landing because the aircraft would not fly in formation during these phases of flight.

[12] American Airlines Newsroom, "Fuel Smart," online, undated; Delta Airlines "Environmental Fact Sheet," online, updated January 2010; JetBlue, *2012 Responsibility Report,* undated (b); United Parcel Service, "Fuel Management and Conservation at the UPS Airlines," online, undated.

[13] JetBlue, 2012.

second includes options the Air Force should explore that have the potential to reduce fuel use in the future. Actions the Air Force should take now include:

- implement engine-out taxiing
- always fly at the optimum flight level and speed
- continue to reduce basic weight of aircraft
- reduce use of the APU
- ensure that loads are properly balanced
- install microvanes on the C-130 fleet.

Additionally, a set of options warrant further analysis and may be able to save an additional 38 MG a year. Further analysis should include:

- expand the use of continuous descent approaches
- continue testing and feasibility studies of vortex surfing
- examine the feasibility of ground towing.

Acknowledgments

We are grateful for the support we received from Air Mobility Command. We particularly appreciate the support and feedback we received from the project sponsors, Brig Gen Scott Goodwin and Maj Gen Scott Hanson.[1] The Fuel Efficiency Office, led by Col Bobby Fowler and Col Keith Boone, was invaluable to our effort. Col Michael Horsey, Maj Darren Loftin, and Maj Daniel Ortwerth continually provided us with the contacts and data we needed for our work. Their enthusiasm for the topic was very much appreciated.

We received a great deal of support from personnel throughout AMC. Maj Mark Blumke, AMC/A4, greatly improved our understanding of maintenance operations. In addition, TSgt Joseph A. Schmeisser assisted us with data collection. AMC's Air Refueling Liaison Office was instrumental in providing data with regards to improving tanker operations. In particular, we are thankful for the support from Mr. Jeffrey Sheppard and Mr. William Ganz. Mr. David Merrill, AMC/A9, provided excellent feedback on our work, which allowed us to sharpen our analysis.

We appreciate the assistance of several industry experts who gave us important insights into commercial aviation best practices. Tom Kane, Ronald Lane, Richard Rolland, and John Dietrich at Atlas provided performance data and explained how they perform detailed flight optimizations. We are grateful for the time they took to expand our understanding of how flight operations can be optimized to minimize fuel use. Art Parra at Federal Express provided insight on the challenges of maintaining and enforcing fuel management procedures. James Barry, William Leber, and Chris Maccarone at PASSUR Aerospace provided details on how information systems, integration of real-time data, and better situational awareness can translate into fuel savings. Tom Randall and Rick Darby at Delta Airlines helped us better understand airline operations and data collection. Quentin Peterson, Michael Swick, Willie Swearengen, Kyle Smith, and C. J. Hybart at Lockheed Martin provided us with a great deal of data on technological modifications to current AMC aircraft. Bill Carolan, Melvin Rice, Mark Stevens, John Dorris, and John Skorupa at Boeing offered perspective on energy efficiency improvements to current aircraft as well as a vision of future technologies.

We are also thankful for the assistance of our RAND colleagues: Michael Kennedy provided expertise and advice throughout the analysis and David Orletsky provided his expert judgment on a wide variety of topics, which helped us shape our research direction. We also appreciate the help of Jerry Sollinger and Karin Suede in the preparation of this document.

We sincerely appreciate the insightful suggestions and observations made by the reviewers of this document, Thomas Light and Tasos Nikoleris.

---

[1] All ranks and offices are current as of the time of this research.

Abbreviations

AFI	Air Force Instruction
AGE	aerospace ground equipment
AMC	Air Mobility Command
APU	auxiliary power unit
ASM	available seat-mile
ATC	air traffic control
BTU	British thermal unit
CDA	continuous descent approach
CDO	continuous descent operation
CG	center of gravity
DLA	Defense Logistics Agency
DoD	Department of Defense
DoE	Department of Energy
EOR	end of the runway
FAA	Federal Aviation Administration
FEO	Fuel Efficiency Office
FL	flight level
FY	fiscal year
GDSS	Global Decision Support System
gph	gallons per hour
ILS	instrument landing system
L/D	lift-to-drag ratio
LDCS	lift distribution control system
MAC	Military Airlift Command
MAF	Mobility Air Forces
MDS	mission design series
MG	millions of gallons

NASA	National Aeronautics and Space Administration
nm	nautical miles
NPV	net present value
NRC	National Research Council
O&S	operations and support
OEW	operational empty weight
OPD	optimized profile descent
PAF	Project AIR FORCE
PB	President's Budget
ppm	pounds per minute
RDT&E	research, development, test, and evaluation
T. O.	Technical Order
ton-nm	ton-nautical mile
UPS	United Parcel Service

1. Introduction

Background

There is significant literature on the topic of reducing aviation fuel use. It is interesting to note that many of the fuel reduction opportunities being pursued today were also encouraged in 1976, when the Military Airlift Command (MAC), now Air Mobility Command (AMC), published a pamphlet entitled "Birds Fly Free, MAC Doesn't."[1] In this document, we will revisit many of those fuel saving concepts presented therein. This illustrates the fact that fuel efficiency is not necessarily about groundbreaking new ideas but rather is more about consistently implementing and following known best practices.

The Mobility Air Forces (MAF) consumes about 60 percent of the Air Force's aviation fuel.[2] As a result of the MAF's relatively large use of aviation fuel, there has been increasing pressure to seek opportunities to reduce this fuel use. The AMC Fuel Efficiency Office (FEO) was chartered in 2008 to identify and implement fuel reduction initiatives.[3] As part of the emphasis on fuel efficiency, the Air Force set a goal to reduce fuel consumption by 10 percent by 2015.[4] In March 2013, the Air Force had already reduced consumption by 12 percent over 2006 consumption.[5] Even though the Air Force has achieved this goal, it is still prudent to pursue cost-effective options to further reduce fuel use.

Literature on fuel use in the aviation industry is extensive; however, there are two difficulties applying these insights to AMC. First, most existing literature does not generally calculate fuel savings at the enterprise level. Second, existing literature focuses primarily on commercial rather than military operations. One primary intention of this work is to apply existing research results at the AMC enterprise level to compare a range of fuel reduction options on an equal basis.

---

[1] Military Airlift Command, Navigation and Performance Division of Aircrew Standardization, "Birds Fly Free, MAC Doesn't," pamphlet, Scott AFB, Ill.: Navigation and Performance Division of Aircrew Standardization, February 10, 1976.

[2] Scott T. Sturkol, "AMC Fuel Efficiency Office Shows How 'Efficiency Promotes Effectiveness,'" Scott Air Force Base, Ill.: Air Mobility Command Public Affairs, January 5, 2011.

[3] Sturkol, 2011.

[4] United States Department of Energy (DoE), *Air Force Achieves Fuel Efficiency Through Industry Best Practices*, Washington, D.C., DOE/GO-102012-3725, December 2012.

[5] Serbu, 2013; Tech. Sgt. Matthew Bates, "Every Drop Counts," Travis Air Force Base, Calif.: Defense Media Activity, November 1, 2013.

Purpose

Our work aims to quantify the fuel savings potential of options for the AMC enterprise. Accordingly, we adapt the methodologies and data found in the literature to AMC operations, making appropriate adjustments in our analysis to account for the differences between military aviation and commercial airlines. We are acutely aware that the Air Force differs in many fundamental ways from a commercial airline, and we make appropriate adjustments in our analysis to account for these differences.

Our research aims to provide the Air Force with

- a methodology to characterize the cost-effectiveness of fuel reduction options
- a prioritization of cost-effective fuel reduction initiatives
- a metric to quantitatively estimate the fuel efficiency of the fleet.

We demonstrate this framework and detailed analysis by comparing a wide array of fuel reduction options in a consistent way that is relevant to AMC. With this approach, we are able to prioritize or rank fuel efficiency options based on their relative cost-effectiveness as well as additional effects of implementation. Finally, we can use these results to derive an estimate of potential fuel consumption reduction for the MAF fleet.

Analysis Approach

This analysis focuses on improvements that can be broadly applied across the fleet. Although we acknowledge the effects of adaptive flying, which vary sortie-by-sortie, we did not include it in our analysis. Analysis of adaptive flying was beyond the scope of this research, partially because the required data were not available.

As an example, consider our analysis of optimum flight levels, where we measure the fuel savings of performing a continuous climb or 2,000-foot step-climbs whenever possible. It is always desirable to climb as the aircraft gets lighter, assuming no winds and standard temperature profiles. However, on any individual sortie, it may be that even when a climb is permissible, it is not beneficial. It may even be the case that making a descent to take advantage of favorable winds would be preferred. These sorts of decisions require a great deal of real-time data and can generally be categorized as adaptive flying and are not considered in our analysis. Additional examples of adaptive flying include derating engines for takeoff, adjusting speed and altitude based on winds and temperature, and flight routing based on winds and traffic.

Key Assumptions

All of the analyses in this document use the following consistent set of assumptions and data unless otherwise noted.

Annual Flight Hours. The total savings potential of any fuel reduction option depends on the baseline fuel use assumed. For more than a decade, the United States has been involved in overseas contingency operations, which has increased the amount of Air Force flying.[6] As the United States continues its withdrawal from Afghanistan, it is expected that MAF flying will decrease. We therefore base our estimates of fuel savings on future projected flying hours rather than on historical flying hours. Specifically, we use the flying hours programmed in fiscal year (FY) 2013 President's Budget (PB) for FY 2014 as an estimation for all future flying. These are reasonable representations of the hours required for crew training and seasoning.[7] Since the programmed flight hours are consistently fewer than historical flight hours of the last decade, this assumption decreases the value of reducing fuel. That is to say, if we instead used historical values, many fuel reduction options would appear more beneficial.

Flight Hour Constraint. Because the FY 2014 programmed flight hours represent the hours required to maintain crew training and seasoning, we assume that these hours represent a minimum. Specifically, if any option exists that reduces flight hours, those flight hours will need to be recaptured on another sortie. This assumption means that this analysis did not consider options that achieve fuel reduction primarily through reducing flight time. Additionally, we did not consider moving training to simulators, improving training efficiency, or using a companion trainer. Although this is a valid assumption if we are measuring fuel use and are constrained by a flying-hour program, it is important to note that they may increase operational effectiveness.

Flying Pattern. We used Global Decision Support System (GDSS) FY 2012 data to derive an estimate of the flying pattern. Specifically, we looked at the distribution of cargo weight and flight distance for each mission design series (MDS). We applied this same flying pattern to future flying and scaled the results such that the total flight hours match the programmed hours.

Dollars and Discount Rate. All dollar figures presented in this document are FY 2013 dollars. We use a 1.1 percent real discount rate throughout this analysis.[8] Although this was the current discount rate as of August 2013, it is the lowest value recorded.[9] By comparison, the discount rate was 3.2 percent in 2003. Today's low discount rate tends to favor investments in new technologies, since fuel savings in the distant future will have greater value today.

---

[6] Christopher A. Mouton, David T. Orletsky, Michael Kennedy, and Fred Timson, *Reducing Long-Term Costs While Preserving a Robust Strategic Airlift Fleet: Options for the Current Fleet and Next-Generation Aircraft*, Santa Monica, Calif.: RAND Corporation, MG-1238-AF, 2013.

[7] Brian G. Chow, *The Peacetime Tempo of Air Mobility Operations: Meeting Demand and Maintaining Readiness*, Santa Monica, Calif.: RAND Corporation, MR-1506-AF, 2003.

[8] United States Office of Management and Budget (OMB), *Guidelines and Discounted Rates for Benefit-Cost Analysis of Federal Programs*, OMB Circular No. A-94 Appendix C, Table of Past Years Discount Rates, revised December 2012.

[9] OMB, 2012.

Fuel Cost. We use the FY 2013 Defense Logistics Agency (DLA) standard price of $3.73 per gallon as our baseline fuel cost.[10] Historically, the DLA standard price has varied significantly. In FY 2012, the standard price was updated three times and ranged from $2.31 to $3.95 in FY 2012 dollars.[11] Although we use the $3.73 per gallon price, our results are presented in a way that accommodates variations on fuel price and other externalities. Extensive research has been done on external costs associated with burning fossil fuels,[12] but inclusion of such cost estimates is beyond the scope of this work. Furthermore, if Energy Information Administration expectations for a rise in real jet fuel prices of 0.7 percent per year through 2040 come true, investments in fuel efficiency will yield greater returns than those estimated in this report.[13]

Flight Modeling

RAND has developed a series of highly adaptable flight models.[14] These models were constructed using flight performance data, such as those found in Portable Flight Planning Software.[15] They allow aircraft weight, specific range, climb performance, and ground fuel usage to be varied dynamically. Because the GDSS data set of FY 2012 missions contains over 100,000 individual sorties, we created sortie bins to allow for faster processing. We created approximately 100 bins for each MDS. We then analyzed the fuel effect of changes to each bin and took the weighted average of these results. The existing suite of flight models did not include the capability to examine optimum speed or altitude flying; we therefore constructed an additional flight model specifically with this capability.

In the case of tankers, which have a significant number of same-base sorties, i.e., sorties where the destination is the same as the origin, GDSS contained insufficient data to model

---

[10] Defense Logistics Agency (DLA), "Standard Fuel Prices in Dollars FY 2013 President's Budget FY 2013 Rates," online, undated (b).

[11] DLA, "DLA Energy Standard Prices," online, undated (a).

[12] Jonathan Koomey, *Comparative Analysis of Monetary Estimates of External Environmental Costs Associated with Combustion of Fossil Fuels*, Berkeley, Calif.: Lawrence Berkeley Laboratory, July 1990; and Interagency Working Group on Social Cost of Carbon, United States Government, *Technical Support Document: Technical Update of the Social Cost of Carbon for Regulatory Impact Analysis Under Executive Order 12866*, May 2013.

[13] Energy Information Administration, *Annual Energy Outlook 2014*, Appendix A – Table 3, undated.

[14] Mouton, 2013; Sean G. Bednarz, Anthony D. Rosello, Shane Tierney, David Cox, Steven C. Isley, Michael Kennedy, Chuck Stelzner, and Fred Timson, *Modernizing the Mobility Air Force for Tomorrow's Air Traffic Management System*, Santa Monica, Calif.: RAND Corporation, MG-1194-AF, 2012; Anthony D. Rosello, Sean Bednarz, David T. Orletsky, Michael Kennedy, Fred Timson, Chuck Stelzner, and Katherine M Calef, *Upgrading the Extender: Which Options Are Cost-Effective for Modernizing the KC-10?* Santa Monica, Calif.: RAND Corporation, TR-901-AF, 2011; Anthony D. Rosello, Sean Bednarz, Michael Kennedy, Chuck Stelzner, F. S. Timson, and David T. Orletsky, *Assessing the Cost-Effectiveness of Modernizing the KC-10 to Meet Global Air Traffic Management Mandates*, Santa Monica, Calif.: RAND Corporation, MG-901-AF, 2009.

[15] FalconView, "Portable Flight Planning Software (PFPS)," online, circa October 2010.

these flights. In these cases, we supplemented the GDSS data set with training flight profiles for the KC-10 and KC-135.[16]

Annualized Costs

The results in this work focus on annualized costs and savings, in terms of both dollars and fuel. We use an indefinite time horizon in our calculations of annualized cost to generate an annual equivalent cost. This methodology is useful when comparing options that have costs or savings that fluctuate over time.[17] Consider, for example, a comparison of two cost-saving options. The first saves $100 for the first ten years and then $10 a year indefinitely. The second saves $125 for 12 years and then has no savings. Using a real discount rate of 1.1 percent, we could determine that the first option is preferred by comparing its net present value (NPV) of $1,776 to the NPV of the second option, which is $1,523.[18]

Although NPV is sufficient for comparing the relative value of options, it is difficult to put the number in perspective. It is therefore usual to convert this to an annualized figure. Specifically, we define the annualized cost or savings to be an infinite stream of a constant amount with the same NPV as the option. In this case, the annualized savings of the first option in our example would be $19.33, that is to say, $100 for ten years followed by $10 a year indefinitely after that has the same NPV as $19.33 a year indefinitely.[19]

Using the same analysis, we can derive an annualized fuel savings. Assuming that in real terms fuel cost remains constant over time, there is a direct conversion between gallons and dollars, and therefore fuel gallons can be annualized the same way as fuel cost.

Fleet Retirements

A key driver in the cost-effectiveness of modifications to the current fleet is the planned retirement schedules. Obviously, aircraft service life predictions contain significant uncertainty, particularly for aircraft with planned retirement dates that are decades off. For example, the KC-135 is estimated to be capable of lasting into the 2040s; however, the

---

[16] Daniel M. Norton, Donald Stevens, Yool Kim, Scott Hardiman, Somi Seong, Fred Timson, John Tonkinson, Duncan Long, Nidhi Kalra, Paul Dreyer, Artur Usanov, Kay Sullivan Faith, Benjamin F. Mundell, and Katherine M Calef, *An Assessment of the Addition of Winglets to the Air Force Tanker Fleets,* Santa Monica, Calif.: RAND Corporation, MG-895-1-AF, January 2012, not available to the general public..

[17] Aswath Damodaran, *Applied Corporate Finance*, 3rd ed., Danvers, Mass.: John Wiley & Sons, Inc., 2011.

[18] The NPV is calculated, as $\text{NPV} = \sum_{k=0}^{\infty} \frac{S_k}{(1+d)^k}$ based on a revenue or savings stream, S_k, where k is the year relative to the current year and d is the discount rate.

[19] The annualized savings of the second option is $16.57. Mathematically, the annualized value, A, of revenue stream, S_k, is $A = d \sum_{k=0}^{\infty} \frac{S_k}{(1+d)^{k+1}}$.

actual likelihood of this is not known.[20] However, given reasonable assumptions, it is difficult to envision a future fleet so radically different from current projections as to alter the conclusions of this work. Figure 1.1 shows our assumed retirement schedule for the current fleet, using estimates from recent RAND work for the airlift fleet[21] and using a KC-46A procurement plan that assumes a one-for-one replacement of the KC-135 at a constant rate until all the tankers are recapitalized.[22]

To illustrate the effect of aircraft retirements on the current fleet, consider a modification to existing C-17 aircraft that saves 4,000 gallons of fuel per aircraft beginning in 2017. This means that through 2032, we will realize 884,000 gallons of fuel savings per year (4,000 × 221 aircraft). The savings will then decrease over time as aircraft are retired, eventually reaching zero in 2058. Calculating the annualized fuel savings as discussed above, we find that an annualized fuel savings of such a modification is 247,463 gallons.

Figure 1.1. Estimated Inventory Profile for the Current MAF Fleet

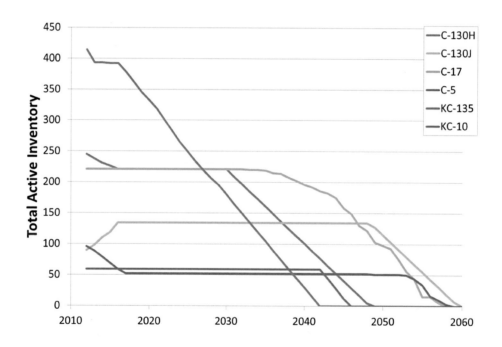

[20] Michael Kennedy, Laura H. Baldwin, Michael Boito, Katherine M. Calef, James Chow, Joan Cornuet, Mel Eisman, Chris Fitzmartin, J.R. Gebman, Elham Ghashghai, Jeff Hagen, Thomas Hamilton, Gregory G. Hildebrandt, Yool Kim, Robert S. Leonard, Rosalind Lewis, Elvira N. Loredo, D. Norton, David T. Orletsky, Harold Scott Perdue, Raymond Pyles, Timothy L. Ramey, Charles Robert Roll, William L. Stanley, John Stillion, F.S. Timson, John Tonkinson, *Analysis of Alternatives (AoA) for KC-135 Recapitalization: Executive Summary*, Santa Monica, Calif.: RAND Corporation, MG-495-AF, 2006.

[21] Bednarz et al., 2012; and Mouton et al., 2013.

[22] Michael Sullivan, Bruce Fairbaim, Keith Hudson, John Krump, May Jo Lewnard, Don Springman, Roxanna Sun, and Robert Swierczek, *KC-46 Tanker Aircraft: Acquisition Plans Have Good Features but Contain Schedule Risk*, Washington, D.C.: Government Accountability Office, GAO-12-366, 2012.

The annualized savings are less than the peak of 884,000 gallons in part because there are no savings at all until 2017 and in part because of the diminishing savings as the fleet is retired.

Comparing the annualized fuel savings to the maximum annual fuel savings, we can develop a fuel savings discount ratio based on the finite life of the MDS. In the case of the C-17, this ratio is 0.28, meaning that only 28 percent of the savings will be realized compared to the case if the C-17s lasted indefinitely. The discount ratios for all of the MDS considered in this analysis are shown in Table 1.1. As the table shows, the discount is greatest (i.e., the discount ratio is lowest) for the KC-135, which will begin retiring with the introduction of the KC-46A, and least for the C-130J and C-5, all of which are projected to remain in the fleet for many years. This discount ratio should not be confused with the financial discount rate.

Table 1.1. Fuel Savings Discount Ratio for MDS Modifications

	MDS Modification Discount
C-130H	0.209
C-130J	0.322
C-17	0.280
C-5	0.332
KC-135	0.119
KC-10	0.248

Cost-to-Carry

Cost-to-carry is defined as the marginal fuel cost of carrying one additional unit of weight one unit of distance. This metric is useful for evaluating the benefits of removing excess weight from an aircraft, in which case the cost-to-carry is avoided. This parameter will be relevant when considering fuel saving options such as reducing the weight of aircraft paint and decreasing the aircraft basic weight. Of course, this is also relevant when analyzing the cost of increasing the cargo onboard a sortie. We estimated the cost-to-carry for the MAF aircraft using RAND-developed flight models and found that across FY 2012 flying, depending on the MDS, the cost-to-carry ranges from 0.018 to 0.036 gallons per ton-nautical mile (ton-nm), as seen in Figure 1.2. The cost-to-carry varies significantly with takeoff weight and flight distance; therefore, for any given sortie, a single averaged cost-to-carry may not accurately capture the cost.

Figure 1.2. Cost-to-Carry for MAF Aircraft Across FY 2012 Flying

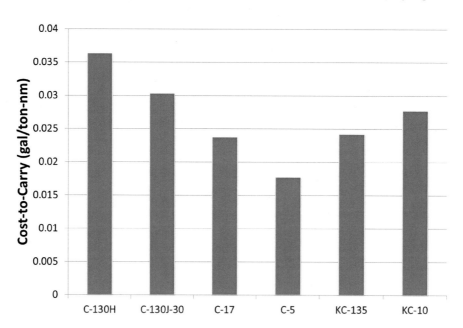

Caveats

Because of the breadth of our analysis, we did not consider sortie-by-sortie variations. Therefore, we do not consider the myriad of limiting factors that may prevent the implementation of fuel savings on any given sortie. Similarly, we do not consider sortie-by-sortie-level differences that can be exploited for additional savings, such as optimizing speed, flight level, and routing to take advantage of winds. We further assume that the fuel required for each sortie is properly calculated; however, as discussed in Chapter Two, this does require investment in information systems. We also do not consider changes to training and seasoning requirements, both of which could lead to significant additional fuel savings. In addition, we assumed a constant real price of jet fuel. Rising jet fuel prices would, of course, make investments in fuel savings more appealing. The Energy Information Administration expects a rise in real jet fuel prices of 0.7 percent per year through 2040.[23]

Observations

Fuel Reduction Potential

A certain amount of fuel is required at minimum to move an aircraft and cargo from one place to another. This amount cannot be reduced without investment in new technology, which, as will be shown below, is generally not a cost-effective proposition for

---

[23] Energy Information Administration, undated.

the Air Force, at least in terms of fuel savings alone. Therefore, although fuel reduction is often spoken of in terms of percentage reduction in total fuel use, large percentage reductions in fuel use can be difficult to achieve in practice. Figure 1.3 shows an approximate breakdown of fuel consumption for C-17s in FY 2012 under nominal conditions. The figure shows that approximately 82 percent of the total fuel consumption is associated with perfect flying given the theoretical limits of the technology. Another 7 percent is associated with ground operations, such as auxiliary power unit (APU) use and taxiing. The remainder, about 11 percent, is associated with imperfect flying. This includes additional flight distance because of weather conditions, airport layout and air traffic control limitations, the requirements to perform step-climbs rather than efficient continuous cruise climb, and a number of other factors.[24]

Our analysis suggests that only 18 percent of the total fuel consumption has even the potential for reduction, absent new technologies or reduced flight hours or cargo loads. Further, much of the "imperfect" flying is unavoidable. Similarly, some, but by no means all, of the ground operations fuel use can be eliminated: Turbofan engines require a warm-up and cool-down period,[25] some APU use is likely required, etc.

We can see, therefore, that a seemingly minor 5 percent reduction in total fuel use actually requires a nearly 30 percent improvement in operations and flying.

Figure 1.3. Approximate Breakdown of C-17 Fuel Consumption

Fuel Reduction Options

After reviewing academic research and existing fuel reduction initiatives in the Air Force and industry, we developed a list of fuel reduction options. Some of these are not likely to be cost-effective for the Air Force, but their inclusion is important to develop the total potential for fuel savings regardless of cost. This list was constructed after reviewing commercial practices and aviation literature, which is discussed in Chapter Two. Below is the list of the options we considered, in the order presented in this document, along with a short description.

---

[24] Federal Aviation Administration, *Instrument Flying Handbook*, Oklahoma City, Okla.: United States Department of Transportation, 2012.

[25] Pratt & Whitney, *Operating Instructions for the PW4000 Series Commercial Turbofan Engines*, online, undated.

Engine-Out Taxiing. Often, not all engines are required for taxiing an aircraft because even the idle thrust produced by a subset of the engines is sufficient to move the aircraft on the ground. By using only the engines required for taxiing, overall fuel use can be reduced.

Optimum Flight Level and Speed. At a given weight, there is an optimum speed and altitude that maximizes an aircraft's fuel efficiency. Adjusting both altitude and speed during flight is therefore necessary to maintain optimum performance.

Basic Weight Reduction. Increased aircraft weight leads to increased fuel burn because of the lift-induced drag. Reducing the weight of an aircraft reduces the fuel required for the flight.

APU Use Reduction. During ground operations, the aircraft's APU can provide power to the aircraft; however, ground units that use less fuel can often provide this power.

Load-Balancing Improvement. The placement of cargo on an aircraft affects the aircraft's center of gravity. As the center of gravity is moved forward, the aircraft becomes less efficient and burns more fuel.

Technical Stop Addition. In addition to the cargo and weight of the aircraft itself, an aircraft must lift the fuel onboard. If, rather than completing a flight nonstop, a technical stop is added, the average fuel weight is reduced and the aircraft burns less fuel in cruise. This savings has to be balanced against considerable logistics and the additional fuel used during the extra landing, stop, and takeoff to determine the cost-effectiveness of this procedure.

Continuous Descent. By performing continuous descents rather than stepping down in altitude blocks, the aircraft can spend more time cruising at fuel efficient altitudes. As a result of the decreased time spent at low altitudes, the aircraft consumes less fuel.

Vortex Surfing. As a result of generating lift, two vortices are generated at each wingtip. It is possible to recover this vortex energy by having another aircraft fly in a particular place in this vortex. By recovering this energy the trailing aircraft is able to burn significantly less fuel.

Paint Reduction. Paint adds to the total weight of the aircraft. If the weight of the paint can be reduced, the fuel burn can also be reduced.

Microvanes. Microvanes are small aerodynamic elements that can be added to the aft portion of a C-130. They reshape the flow around the aft cargo door and reduce the total drag on the aircraft, thereby reducing fuel use.

Ground Towing. Aircraft engines are particularly inefficient when operating at low speeds on the ground. As a result it is much more efficient to tow the aircraft rather than have them taxi under their own power. Although this requires additional logistics, it does offer significant fuel savings.

Lift Distribution Control. By actively controlling the distribution of lift along the wing, it is possible to change the angle of attack at which the aircraft flies. For the C-130, this serves to reduce the drag, in much the same way as microvanes.

fuel reduction potentials for a wide array of conservation measures in a commercial context. We also conducted interviews and engaged in on-site visits with experts at leading U.S. passenger airlines, cargo carriers, air traffic management firms, and aircraft manufacturers (e.g., FedEx, Delta, Passur, Boeing, Lockheed Martin, and Atlas Air).

Differences Between Commercial Aviation and the Air Force

It is important to acknowledge fundamental differences between the business model of a commercial airline and the operating model of the U.S. Air Force, which limits the relevance of some commercial best practices to the Air Force. For example, commercial airlines operate in congested airports with the number of departing and arriving aircraft often exceeding the maximum limit the airport can efficiently handle at a given time. This leads to taxi queues, off-gate holds, and other disruptive events that will lead to excessive fuel consumption. These occurrences are much less common on Air Force bases. Consequently, it makes sense for commercial aviation but not military aviation to invest in surface congestion management systems. Furthermore, the ground infrastructure available to commercial aviation is starkly different from military infrastructure. Passenger airlines use well-developed gates at almost every major civilian airport. These gates consistently provide ground power and air-conditioning units, so that APU usage on passenger airlines is minimized. In contrast, the Air Force is charged with operating from bases that often may not have the amenities of a major international airport. Most important, the high demand for commercial air transportation also results in a much higher utilization rate of aircraft by commercial carriers. In general, commercial airlines fly their aircraft more often and replace them more frequently. Each replacement provides the opportunity to capture new technological improvements in fuel efficiency. High utilization also changes the cost-effectiveness of many technological conservation measures with up-front capital costs, such as winglets and aerodynamic improvements. In a generalized analysis of the optimal time between new aircraft programs, we see important differences in cost structure between commercial and military aircraft. The total cost of acquiring, operating, and maintaining a fleet of aircraft includes research and development, procurement, operating and support, crew costs, and fuel costs. Research and development costs for commercial aircraft are included in the purchase price of the aircraft, effectively spreading the cost across the multiple airline customers, but for military aircraft, these costs are borne entirely by the military. Moreover, with multiple airline customers for a commercial aircraft, total purchase sizes are much greater for commercial carriers than for the military. Because of economies of scale and manufacturing efficiencies over time, the per-unit cost of a representative commercial aircraft and a comparable military aircraft would be less in this simple case. Thus, from a purely cost-minimizing point of view, commercial airlines have a greater incentive to upgrade to new generation of aircraft (and concurrently capture technological improvements in fuel efficiency) more frequently.

Technological Improvements

Jet aircraft today are 70 percent more fuel efficient on a per passenger-mile basis than they were in the 1960s because of improvements in engines and airframe design.[2] Unfortunately, technological diffusion is a long process. It can take up to a decade to design an aircraft, and then production can run over 20 to 30 years with each aircraft having a service life of 25 to 40 years.[3] Improvements occur at the pace at which new products are introduced into the fleet.

Often, progress in different technologies occurs concurrently. Although every successive generation of aircraft contains multiple improvements, the most significant technological improvement areas are

- improvements in the performance of the engine to reduce fuel burn per unit of delivered thrust
- improvements in aerodynamics to reduce drag
- reductions of aircraft weight.

Propulsion

Improvements to engine performance contributed to a majority of total fuel efficiency gains in aviation. One study estimated that 57 percent of reductions in aircraft fuel consumption between 1959 and 1995 resulted from engine improvements that increased engine pressure, temperatures, and the engine bypass ratio.[4] Improvements in engine efficiency have the potential to result in positive "snowballing" spillovers, because reduced consumption allows for smaller fuel tanks, which leads to lighter aircraft, which in turn results in even lower fuel consumption. Therefore, a new engine-airframe design, where the airframe is optimized to the new engine, is always superior in terms of fuel efficiency. Nevertheless, it is still possible to implement retrofit solutions for existing aircraft by either re-engining or replacing the engine module with a more efficient technology version. However, these measures do not come cheaply—the typical cost of replacing engines on a

---

[2] United States Government Accountability Office, *Aviation and Climate Change*, Washington, D.C., GAO-09-554, 2009; and International Civil Aviation Organization, *ICAO Environmental Report 2010: Aviation and Climate Change*, 2010 (b).

[3] ICAO, 2010 (b).

[4] R. Babikian, S. P. Lukachko, and I. A. Waitz, "The Historical Fuel Efficiency Characteristics of Regional Aircraft from Technological, Operational, and Cost Perspectives," *Journal of Air Transport Management*, Vol. 8, No. 6, 2002, pp. 389–400.

ten-year-old airframe could cost $14 million to $62 million, depending on the aircraft and engine.[5]

Aerodynamics

Lift-induced drag and skin friction drag are the greatest contributors to aerodynamic drag. Winglets are particularly appealing for reducing lift-induced drag, particularly when airport and hangar or gate sizes limit wing spans. Alaska Airlines' entire fleet of 737s has winglets, saving 100,000 gallons of fuel per aircraft per year.[6] American Airlines first retrofitted winglets on its 737 and 757 aircraft in 2005. The projected annual fuel savings is about $39 million. Delta Airlines installed winglets on all its 737s and a fraction of its 767s, 757s, 747s, and A330s, and future purchases of 737s and Airbus narrow bodies will all come with winglets.[7]

Varying camber involves altering lift characteristics through leading and trailing edge devices on the wing. Boeing and the National Aeronautics and Space Administration (NASA) are testing a Variable Camber Continuous Trailing Edge Flap system, which would provide both high lift for takeoff and landing and reduced drag during cruise, through active control of 42 individual flap sections to change the twist of a flexible wing.[8]

Reducing friction drag is thought to offer the greatest potential for aerodynamic efficiency improvements over the next ten to 20 years.[9] Table 2.1 presents the array of aerodynamic technologies considered by the International Coordinating Council of Aerospace Industries Associations for the 2010 Fuel Burn Technology Review. A promising, but limited, near-term approach is reducing excrescences, which include any surface imperfections or roughness that results from manufacturing (e.g., seals, rivets) as well as protuberances, such as antennas and air inlet/exhaust devices. However, the potential for improvement in lift-to-drag ratio (L/D) is only about 1 percent. A less-mature technology that promises more substantial improvements is reducing skin friction by maintaining laminar flow by means of a natural laminar flow and hybrid laminar flow control. These two technologies have been demonstrated in flight tests of wings, nacelles, empennages, and winglets on a Boeing 757, Dassault Falcon 900, and Airbus A320.[10]

---

[5] Pamela Farries and Chris Eyers, *Aviation CO2 Emissions Abatement Potential from Technology Innovation*, London: Committee on Climate Change, QINETIQ/CON/AP/CR0801111, October 14, 2008.

[6] 100,000 gallons per aircraft per year results in an annual fleetwide savings of 2.2 million gallons (MG), constituting 10 percent of total savings from its fuel savings program.

[7] Elyse Moody, "Focus on Fuel Savings," *Aviation Week & Space Technology*, March 1, 2012.

[8] Graham Warwick, "NASA, Boeing Study Flexible Wing Control," *Aviation Week & Space Technology*, January 23, 2013 (a).

[9] ICAO, 2010 (b).

[10] ICAO, 2010 (b).

Table 2.1. Friction Drag Reduction Technologies

	Technology Maturity	L/D Improvement (percent)
Riblets	Low/Med	1 to 2
Natural Laminar Flow	Med	5 to 10
Hybrid Laminar Flow Control	Low/Med	5 to 10+
Excrescence Reduction	High	1
Variable Camber	Med/High	2

SOURCE: ICAO, 2010 (b).

Similarly, riblets consist of a thin grooved layer of plastics applied to air-swept surfaces. Inspired by the tiny ridge-like structures on shark skin, NASA in conjunction with 3M developed a thin adhesive film and flight tested it on a Learjet in 1986, achieving an 8 percent drag reduction. In 1989, Airbus flight-tested an A320 with the film on 75 percent of its surface and found less than a 2 percent reduction in fuel burn.[11] Cathay Pacific was the only airline to use riblets, applying them to 30 percent of an A340-300 used on long-haul flights, but found that the lifespan of the film was only two to three years.[12] After the 1980s, interest in riblets faded, although recent innovations have surfaced to allow embedding the riblet pattern into the aircraft paint to aid durability.[13]

In practice, airlines simultaneously invest in multiple aerodynamic improvements. We see this notably for long-range 777s, with United, Continental, and Delta investing in modifying the outboard aileron droop and the environmental control system, making ram air improvements, and installing smaller wing vortex generators to yield a total of 1 percent fuel burn improvement.

Weight Reduction

With each successive generation, manufacturers have reduced weight through progressive introduction of advanced alloys and composite materials, improved and new manufacturing processes, and new systems. For example, the Boeing 777, introduced in 1995, uses lighter-weight composite material for 12 percent of the airframe (by weight),

---

[11] J. Szodruch, "Viscous Drag Reduction on Transport Aircraft," AIAA Paper 91-0685, 29th Aerospace Sciences Meeting, Reno, Nev., January 7–10, 1991.

[12] Clyde Warsop, "Current Status and Prospects for Turbulent Flow Control, Aerodynamic Drag Reduction Technologies," in Peter Thiede, ed., *Proceedings of the CEAS/DragNet European Drag Reduction Conference,* June 19–21, 2000, Potsdam, Germany.

[13] Lufthansa Technik, "Like a Shark in the Water: Innovative Lacquer System to Reduce Drag," online, October 2012.

and the 787, introduced in 2011, uses 50 percent composites. Advanced welding technologies remove the need for traditional rivets, simultaneously reducing weight, lowering manufacturing costs, and reducing drag. New systems, such as fly-by wire, reduce the need for flight control cables and reduce the size of vertical and horizontal stabilizers, thereby reducing weight.

Airlines have also taken aggressive steps to reduce cabin weight by removing unnecessary items, such as old phone equipment, galley tables, magazine racks, razor outlets, and logo lights; minimizing supplies of necessary items carried on board, such as water; and replacing old, heavy cargo containers and catering carts with newer, lighter models. American Airlines claims that its electronic flight bag program, in which Apple iPads replace 35-pound pilot kitbags, saves 400,000 gallons of jet fuel annually.[14] American Airlines is also notable for having previously not painted its airframes to reduce weight and instead using polish to combat corrosion. However, the fuel saved from reduced weight is partially offset by the increased maintenance required for polished aircraft. JetBlue began a weight reduction campaign in 2006, removing hundreds of pounds from each aircraft. For example, changing to radial tires on the E190 saved 128 pounds per plane. These modifications, in addition to changes to galley provisioning and galley cart weights, are claimed to have saved over 950,000 gallons of fuel in 2007.[15] For commercial airlines seeking to reduce weight, it seems that no item is too small to escape notice: British Airways reportedly sells inflight alcohol in plastic bottles instead of glass to cut down on weight.[16]

Another way to reduce weight is to carry less fuel. Alaska Airlines carefully counts the number of children onboard and reduces the amount of fuel loaded if children fly instead of adults.[17] American Airlines culls past flight data to better calibrate the amount of fuel carried to what has been historically consumed on a route based on aircraft type, traffic, and weather conditions. American also monitors ground crews regularly to ensure that they pump the precise amount of fuel ordered.[18] American and Continental won approval from the Federal Aviation Administration (FAA) to halve the amount of reserve fuel that must be carried on some international flights, arguing that the legacy 10 percent reserve requirement originating in the 1950s was outdated given modern navigation systems:

---

[14] American Airlines Newsroom, "Fuel Smart," online, undated.

[15] JetBlue, *2007 Environmental and Social Report*, online, undated (a).

[16] Joseph C. Anselmo, "Fuel Crisis Forces Airlines to Conserve, Drop by Drop," *Aviation Week & Space Technology*, December 6, 2004, p. 54.

[17] Anselmo, 2004, p. 54.

[18] J. Hilkevitch and J. Johnsson, "American Airlines' Effort to Cut Fuel Reserves Draws Fire from Pilots," *Los Angeles Times*, July 13, 2010.

American anticipates an $11 million annual savings from this change.[19] Other airlines, such as Atlas Air, reduce the international reserve fuel carried by using a redispatch technique: filing a flight plan to an intermediate airport that is closer than the intended destination and then re-dispatching to the intended final destination if fuel consumption checkpoints are cleared.[20]

Operational Improvements

Operations encompass a broad range of activities including on-the-ground airport activities, airborne activities (e.g., flying the aircraft, air traffic management), and fleet management activities. "Gate-to-gate" efficiency opportunities exist at every stage of flight, beginning with planning activities before passengers or cargo are loaded and ending after arrival in the terminal. Operational measures may be particularly beneficial in that they save fuel and do not necessarily require the introduction of new expensive technology. We have organized commercial operational improvements into three categories: ground efficiencies, airborne efficiencies, and fleet efficiencies.

Ground Efficiencies

The first mitigation opportunity occurs when the aircraft is still parked at the terminal gate. One conservation measure is to limit the use of fuel intensive APUs and instead use more efficient ground equipment. Although new engines have achieved steady gains in fuel efficiency, the rate of improvement in APU fuel consumption has lagged behind.[21] JetBlue has worked with management at airport gates and maintenance hangars to ensure the availability of ground power and preconditioned air: Of the 177 gates it regularly serves, 129 have ground power and 115 have preconditioned air.[22] American anticipates some of its greatest year-over-year savings from this measure, more than 6 MG a year.[23] In 2011, United and Continental achieved ten-minute and five-minute average reductions, respectively. During interviews with industry, we found that one major carrier was able to reduce APU usage below 30 minutes on a significant majority of its flights.

Once the aircraft pushes back from the gate, several measures address the need to move the aircraft to the runway efficiently. Engine-out taxiing involves shutting down one or more engines during taxiing. JetBlue pilots and technical operations crewmembers

---

[19] Anselmo, 2004, p. 54.

[20] Steve Altus, "Effective Flight Plans Can Help Airlines Economize," *Aero Quarterly* (Boeing), Vol. 3, 2009, pp. 27–30.

[21] Air Transport Action Group, "Beginner's Guide to Aviation Efficiency," November 2010.

[22] Moody, 2012.

[23] American Airlines Newsroom, undated.

routinely use single-engine taxi procedures. Their E190 fleet has used single-engine taxi since its beginning in 2005, and single-engine taxiing is standard operating procedure for their A320 aircraft, unless weather and airport layout cause it to be infeasible. At large airports with typically long taxiing times, applying this measure could be beneficial. Overall, it is implemented on over half of all JetBlue flights, and at JFK International Airport, where significant delays and congestion are often experienced; it is used on 88 percent of A320 flights and 95 percent of E190 flights in 2007.[24] Alaska Airlines practices this measure systemwide and claims to save 260,000 gallons of fuel per year, according to 2009 figures.[25]

An alternative to taxiing altogether is to use aircraft tow tractors to move aircraft longer distances rather than using the aircraft's own, less-efficient engines. This technique also allows the delay of engine start until about five minutes before departure. American Airlines uses tow tractors to reposition aircraft between terminal gates and line maintenance hangars, saving more than 4 MG of fuel in 2011.[26]

Another area of improvement is reducing bottlenecks and congestion on the ground by improving airport operations. Especially at airline hubs, a large number of aircraft arrive and depart within a narrow window of time so that passengers can make connecting flights. This often results in traffic jams on the runway, which waste fuel as airplanes sit idling. To compound this problem, air traffic control generally implements a "first come, first served" system, allowing any aircraft to push off into extended taxiing queues and ignoring prioritization of high-value flights.[27] The result of surface congestions is increased taxiing times and fuel burn. In 2009, there were over 32 million minutes of departure taxiing delay at major U.S. airports, which translates to 130 MG of excess fuel burn.[28] Recently, major airlines have attempted ad hoc refinements at their respective hubs: Delta reoriented traffic in Atlanta, and US Airways asked the FAA for approval to use a shorter, less-frequented runway at its Philadelphia hub to better choreograph departures.[29] Taking a more systematic solution, PASSUR Aerospace is working to deploy "departure metering" in collaboration with most major airlines and over 50 airports.[30] The concept of departure metering is to determine the number of departing and arriving aircraft an airport can

---

[24] JetBlue, undated (a).

[25] Moody, 2012.

[26] Moody, 2012.

[27] PASSUR Aerospace, "Arrival Management," fact sheet, online, undated.

[28] Alex Nakahara, Tom G. Reynolds, Thomas White, Chris Maccarone, and Ron Dunsky, "Analysis of a Surface Congestion Management Technique at New York JFK Airport," paper prepared for 11th AIAA Aviation Technology, Integration and Operations (ATIO) Conference, Virginia Beach, Va., September, 2011.

[29] Anselmo, 2004, p. 54.

[30] PASSUR Aerospace, "PASSUR Aerospace Reports Revenue Increase of 22% for Fiscal Year 2010," Stamford, Conn., January 31, 2011.

efficiently handle, spread out flights more evenly throughout the day below this efficiency limit, and prioritize pushbacks based on quotas allocated to individual airlines. Excess flights are redistributed to later time intervals when they can be more efficiently accommodated. In 2010, a full-time implementation of prototype software and processes was put in place by PASSUR for the Port Authority at JFK International Airport to manage disruption caused by a five-month closure of its longest runway for maintenance. An estimate of the annualized monetized benefits of departure metering at JFK, calculated by the Massachusetts Institute of Technology, was a 14,800-hour reduction in taxiing times and a savings of 5 MG of fuel.[31]

Airlines have also prioritized maintenance activities such that fuel efficiency is second only to safety. A malfunctioning air conditioning pack is an example of an issue that significantly affects fuel efficiency and would be considered a priority, even though the airplane could still be operated comfortably and safely. If the air condition pack malfunctions, the aircraft must travel at a lower, suboptimal altitude, resulting in slower speed and increased fuel burn.

Both Atlas and American report carefully tracking the performance of each aircraft, inspecting the ones that underperform, and identifying issues and making repairs. Such efforts uncover even seemingly insignificant flaws, such as a tiny dent half the size of a dime on the leading edge of an MD-80 wing, which was reducing efficiency.[32] This also allows for more precise calculations of fuel requirements and helps ensure crew confidence in these calculations.

Another maintenance strategy to save fuel is instituting more frequent engine washing procedures. Calibrated nozzles direct heated, pressurized water into the inlet and through the gas path while the engine is spinning to remove deposit buildup from the blades, lowering exhaust gas temperatures as well as increasing performance and fuel efficiency. Delta previously washed engines as needed in response to poor exhaust gas temperature margins, but it now washes all its engines with Delta-manufactured equipment at specified intervals with few exceptions. During 2011, Delta washed 1,261 Delta engines and 33 contract engines, to which it attributed $1.6 million in fuel cost savings.[33] American began its program in 2007 with maintenance staff at six primary sites and four auxiliary sites washing American's engines twice a year. In 2011, it claimed 7.3 MG of fuel saved.[34] JetBlue and United estimated a savings of 1 MG and 3 MG per year, respectively.[35] Although some of the engine washing for United occurs off-site, almost 90 percent of

---

[31] Nakahara et al., 2011.

[32] Anselmo, 2004, p. 54.

[33] Moody, 2012.

[34] Moody, 2012.

[35] Moody, 2012.

United's engine washing is done on-site, and the airline sees much room to optimize the washing program in terms of the frequency, location, and alignment of equipment to better serve its wide network.

Airborne Efficiencies

A common fuel conservation strategy is to use a cost index to optimize flying behavior in response to fuel prices and other flying costs. The cost index defines the relationship between the time-related cost of airplane operation and the cost of fuel. Specifically, the cost index is a value within an allowable range (e.g., 0–9,999 for the 747-400) that is entered into the flight management system:[36]

$$\text{Cost Index} = \frac{\text{time cost (\$ per hr)}}{\text{fuel cost (cents per lb)}}$$

The time-related direct operating cost excludes the cost of fuel but includes flight crew wages, hourly maintenance costs, and the associated hourly costs of owning and operating engines and APUs. Changing the cost index in favor of reduced fuel consumption will adjust the flight speed and how the aircraft is programmed to climb and descend. A low cost index is used when fuel cost is high relative to other operating costs. If the cost index is zero, then the minimum fuel flight is preferred. Conversely, if the cost index is very high, then a fast flight profile will be preferred. Table 2.2 shows an evaluation conducted by Boeing for an airline to find the optimal cost index for its 737 and MD-80 fleet. It also projects the time effect and the annual cost savings of flying the optimum over a typical 1,000-mile trip. We can see significant annual cost savings with a negligible effect on schedule (less than or equal to three minutes). Likewise, JetBlue decided to reduce speed, adding two minutes to each A320 flight but gaining savings of 188 pounds of fuel per flight.[37]

Table 2.2. Cost Index Impact

	Current Cost Index	Optimum Cost Index	Time Impact Minutes	Annual Cost Savings ($M)
737-400	30	12	+1	0.75-0.77
737-700	45	12	+3	1.79-1.97
MD-80	40	22	+2	0.32-0.43

SOURCE: Roberson and Pilot, 2007.

---

[36] B. Roberson and S. S. Pilot, "Fuel Conservation Strategies: Cost Index Explained," *Aero Quarterly* (Boeing), Vol. 2, 2007, pp. 26–28.

[37] JetBlue, undated (a).

According to the FAA, NextGen, the planned air traffic management system built for increased aviation traffic in the future, will indirectly reduce fuel consumption by reducing congestion and enabling optimal routes.[38] NextGen consists of several components that integrate technological and operation improvements. First, it offers improved navigational capability and precision, which allow for more precise control during flight approach and descent so that aircraft fly as close as possible to their preferred 4-D trajectory. The concept of 4-D trajectories is particularly important in NextGen because aircraft will need to be at a particular point in space at a particular time. Second, NextGen network-enabled weather will provide advanced real-time weather data to help reduce delays caused by weather and allow aircraft to best use wind conditions to improve efficiency. Third, continuous descent approaches (CDAs) enable planes to perform more fuel efficient approaches and are already in place in a number of U.S. airports. United Parcel Service (UPS) first began testing CDAs on flights to its Louisville facility in 2008. Over a few months, the average savings was 50 gallons of fuel per flight.[39] In 2009, Georgia Tech's Air Transportation Laboratory conducted a study at Atlanta's Hartsfield-Jackson International Airport in collaboration with the FAA, FedEx, Delta, and AirTran to develop CDA procedures for a complex airspace with varying aircraft types, weights, wind conditions, and airport configurations. Currently, CDA has been demonstrated in over 60,000 landings at Los Angeles International Airport. A similar upgrade of air traffic management is currently ongoing in Europe through the Single European Sky Air Traffic Management Research Program.

Fleet Efficiencies

It is well understood that retiring fuel inefficient aircraft and replacing them with more fuel efficient models can achieve additional fuel savings. Figure 2.2 shows variations in fuel consumption, expressed as gallons per available seat-mile (ASM), within and between aircraft classes.[40] In 2008, American and Alaska replaced many of its relatively inefficient MD-80 aircraft with more efficient Boeing 737-800 aircraft, which consume 18 percent less fuel.[41] The same year, Continental replaced regional jets on many of its routes with turboprop planes (regional jets are on average less fuel efficient than turboprops of the same seat size).[42] In general, operating a newer fleet with the latest technologies means less

---

[38] GAO, 2009.

[39] Alex Kingsbury, "New Landings Save Airplane Fuel," *U.S. News and World Report*, July 2, 2008.

[40] J. Morrison, P. Bonnefoy, R. J. Hansman, and S. Sgouridis, "Investigation of the Impacts of Effective Fuel Cost Increase on the U.S. Air Transportation Network and Fleet," *10th AIAA Aviation Technology, Integration, and Operations (ATIO) Conference*, Fort Worth, Tex., September 2010.

[41] Moody, 2012.

[42] Babikian, Lukachko, and Waitz, 2002.

Figure 2.2. Aircraft Fuel Intensity, by Class

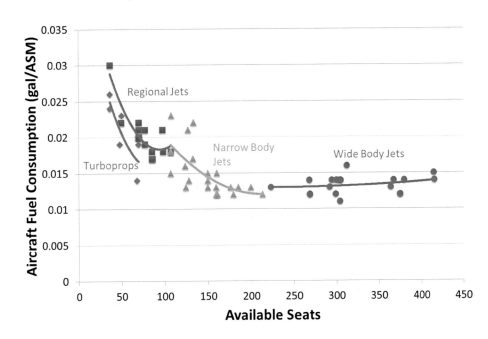

SOURCE: Morrison et al., 2010.

fuel consumption and reduced maintenance costs. Delta is planning a turnover of its aging fleet by removing 130 of its least efficient aircraft by the middle of 2013, including its entire DC-9 fleet, phasing out Saab 340 turboprops, and reducing the number of 50-seat regional jets.[43] United expects to acquire 132 new aircraft between now and 2019, consisting mainly of Boeing 787s, 737s, and Airbus A350s. These new aircraft will result in a 20 percent efficiency increase.[44]

The academic economics literature has also studied three main ways in which airlines conduct business operations in response to increasing fuel price by employing theoretical models of airline competition.[45] First and most obviously, airlines attempt to pass the burden of fuel costs onto the customer in the form of increased airfare or fuel surcharges. Although the U.S. carriers operating in a hyper-competitive market have had difficulty increasing their prices, many foreign carriers such as British Airways, Air France-KLM, and Qantas have been successful in imposing significant fuel charges of up to $20 per flight. Air France-KLM claims that its surcharges have paid 50 percent of the increased

---

[43] The current average age of Delta's 722 aircraft is 16 years, in contrast with budget competitors Horizon and JetBlue with average ages of five to six years.

[44] Moody, 2012.

[45] J. K. Brueckner and A. Zhang, "Airline Emission Charges: Effects on Airfares, Service Quality, and Aircraft Design," *Transportation Research Part B: Methodological*, Vol. 44, Issues 8–9, September–November 2010, pp. 960–971.

fuel bill.[46] If increasing airfare fails, a second coping strategy is to alter the quality of the service provided, such as reducing flight frequencies, decreasing seat room, and increasing passenger load factors (i.e., the percentage of seats occupied on a flight). Between 1995 and 2008, U.S. airlines increased their load factor from 65 percent to 80 percent, effectively increasing fuel efficiency on a per-passenger basis.[47] A third and more drastic business adaption would be a fundamental change in the structure of airline networks. Theoretically, an airline could operate either a hub-and-spoke network or a fully connected, point-to-point network, and each has a differing level of fuel efficiency. A sufficient increase in fuel price could generate a movement from current hub-and-spoke networks, which capture economies of aircraft size, toward fully connected, point-to-point networks, which reduce the total distance flown for passengers connecting through hubs.

Finally, airlines have become increasingly creative in expanding their business outside traditional airline operations in attempts to curtail fuel costs. Most major passenger carriers, with the exception of US Airways, purchase financial instruments to receive fuel in the future at a certain price to hedge against volatile prices.[48] The benefit is increased time to adapt to true market prices in the long run without being subjected to shocks in the short run. From 1998 through 2008, Southwest implemented aggressive fuel hedging by securing about 80 percent of its annual fuel in advance, at a cost of $25 million in one year, and saved $3.5 billion over what it would have spent if it had paid the average price for jet fuel during that time.[49] Early on, most other airlines made the mistake of not following suit, confident either that they could pass on any increases in price to the customer or that the price would eventually fall. However, hedging does not always save money, and it can be argued that focusing on short-term costs by spending money on hedging contracts comes at the expense of long-term investments in fuel management and efficiency. One unconventional long-term strategy was demonstrated by Delta, which was the first U.S. carrier to purchase its own jet fuel refinery in 2012 to reduce exposure to fuel price risk.

Implementation

Having identified a wide array of technological and operational improvements, we observe that putting these measures into practice is often associated with a sophisticated integration of information systems, coordination across multiple departments within the airline, and consistent efforts to sustain results achieved. Fuel efficiency programs are

---

[46] Anselmo, 2004.

[47] GAO, 2009.

[48] S. Carey, S. Chaudhuri, and T. Stynes, "Delta, US Airways Strike Positive Note," *Wall Street Journal*, October 24, 2012.

[49] Dan Reed, "Can Fuel Hedges Keep Southwest in the Money?" *USA TODAY*, online, updated July 24, 2008.

hardly ever "one size fits all." Different airlines tend to choose different focuses. The merger of United and Continental in 2010 allowed a unique opportunity for those within both companies to compare historical best practices. Continental had primarily emphasized technological solutions, such as aircraft modification, fleet replacement, and the installation of sophisticated data-collection devices in its fuel efficiency efforts. Meanwhile, United's relative strength was in refining operational procedures and planning and executing optimal flight policy.[50] The postmerger fuel efficiency program capitalized on the individual strengths through a new fuel management information system that combined each company's technological capability of real-time data collection and reporting with the operational know-how associated with analyzing data to make continuous improvements.

Information System

Information systems help flight crews and planners make decisions that minimize the total cost of a flight. There are four common components to a state-of-the-art information system:

- integrated surveillance network
- performance database
- predictive analytics
- decision support tools.

Finding new fuel saving opportunities means first gathering actionable, high-quality information. This is normally accomplished through an integrated surveillance network that automatically records performance measurements and reports data collected regularly (and ideally, in real-time for quick feedback). Automatic collection, as opposed to self-reported or ad hoc collection, encourages data consistency and reliability. For example, FedEx records more than 150 data points per flight from multiple systems.[51] PASSUR has a unique integrated surveillance network that ingests government and partner airline data feeds combined with their own surveillance sensor network to pull in live air traffic information transmitted from other aircraft. Its network compiles extensive amounts of key metrics, such as arrival rates, delay minutes, diversions, frequency and duration of gate conflicts, luggage and passenger misconnects, ground crew workforce assignments, and on-time departures.

Years of historical flight information can be mined for trends at all levels (e.g., routes, airframes, stations, and crews) to discover cases of underperformance. For example, Atlas Air regularly employs data validation by comparing current to average historical performance to identify stations where excessive fuel is loaded before takeoff or airframes

---

[50] Moody, 2012.

[51] Private communication with FedEx, March 26, 2013.

that need flight-control tuning. Southwest recently upgraded its databases to reconcile consumption and purchase information on a daily basis.[52]

Of particular interest for airlines is the ability to leverage past performance data to generate accurate predictions of fuel requirements for a particular aircraft to complete a particular route.

Finally, many flight-related decisions have complex spillover effects to the network, and crews and planners need tools that enable them to compare options. Current decision support tools include flight planning software that calculates the most efficient route and speed given weather and wind predictions, terrain, fuel prices, flyover fees, etc. UPS recently implemented a system called Lufthansa Systems Lido Operations Center.[53] Future support tools will also display anticipated air traffic congestion, real-time arrival airport delays, and other airspace complexities, allowing the pilot to respond dynamically by adapting flight speed and path while en route to save fuel. Such decision support tools provide crews and planners with increased visibility and continually remind them of specific fuel effects of their actions.

Integrating Fuel Conservation into Company Culture

Once fuel conservation objectives have been identified, an equally important task is embedding fuel conservation into company culture and training. In our survey of commercial practices, we find that putting fuel saving operations as an overlay onto standard operations limits the effectiveness. Fuel economy management is a cross-departmental effort for many major airlines. Alaska Airline has a fuel conservation steering committee that is jointly chaired by the vice president of maintenance and engineering and the vice president of flight operations. Delta relies on collaboration between its flight operations, environment and safety, and fleet groups to track fuel efficiency. United's fuel efficiency program also reports monthly on the carrier's portfolio of initiatives to a fuel council composed of representatives with expertise in operational engineering, load planning, dispatch technology, and flight dispatch support.[54] American's company-wide effort is called "Fuel Smart" and publishes historical results achieved from 2005–2011.[55]

Atlas Air's FuelWise program extends across all departments, including maintenance, ground operations, flight operations, information technology , and system control. At Atlas's training center, the authors observed that conservation measures from FuelWise are built into standard operating procedures and reinforced through training and simulator events. From the start, the crew publications and flight operations manuals espouse

---

[52] Anselmo, 2004.

[53] United Parcel Service, "Fuel Management and Conservation at the UPS Airlines," online, undated..

[54] Moody, 2012.

[55] American Airlines Newsroom, undated.

procedures that emphasize fuel conservation. This includes delayed engine start, taxiing on less than all engines, reduced takeoff thrust, optimum flaps, optimized altitude and routing, precision navigation, optimum (idle power) decent, minimal use of reversers, and delayed APU start before departure and on arrival. The exact takeoff thrust and flap settings come from performance calculations where consideration of fuel conservation is the default criteria.

Atlas uses an information system to record data, identify potential savings areas, and provide decision support. This allows performance calculations to be constantly refined with new data and analysis. The information system is also vital to creating a constant feedback loop between flight operations and the crew. High-granularity data tracking of fuel consumption by flight, captain, tail number, and dispatcher enables validation of procedures and accountability. By constantly monitoring flight performance, causes of inefficiencies can be immediately addressed. If a pilot is identified as not adhering to standard procedure, feedback and sensitization can be provided quickly, with the option for retraining. Monitoring and reinforcement also occur within a crew. The result is a culture where efficient operations are the standard, eliminating the impression that saving fuel is another task.

Conclusion

In this chapter, we reviewed a wide range of commercial best practices for fuel conservation, grouped by technological and operational improvements. Technological advances in propulsion, aerodynamics, and weight reduction have contributed strongly to progress in efficiency since the introduction of jet aircraft. Many of the technological upgrades to engines and airframe occur over longer time horizons, but changes to weight and aerodynamics have been actively pursued by airlines in the short term. Although the percentage improvements from operational opportunities are relatively small, the sheer volume of flying done by commercial airlines yields significant absolute fuel savings. We identified opportunities for fuel conservation at every stage of a flight: engine start, taxi-out, takeoff, climb, cruise, descent, landing, taxi-in, and engine shut-down. For each conservation measure, airlines and academic sources offer disparate estimates of potential and realized benefits. Unfortunately, these estimates are given across different periods of time, during which fuel prices varied widely. Also, different airlines have different sizes and scales of operation, so that comparability of absolute savings across airlines and across time provides little insight into the relative effectiveness of different conservation strategies.

Finally, we characterized fundamental differences between the business model of a commercial airline and the operating model of the U.S. Air Force. These differences affect incentives to invest in both information systems and technological upgrades to aircraft, requiring significant up-front capital expenditures. Certain information systems provide detailed, precise visibility into airline operations, and the high level of commercial utilization justifies their high costs. The challenge for the Air Force in adapting commercial best practices is determining not only whether these solutions are cost-effective but whether they can be similarly replicated with low-cost substitutes that are more appropriate for the Air Force operating model. For example, rather than installing expensive sensors in all aircraft to record and transmit APU usage, the Air Force could intensify monitoring and auditing of APU use as part of a quality assurance or other compliance inspection. The rest of this report is focused on adapting and evaluating many of the ideas in this chapter to the specific context of the U.S. Air Force.

3. Cost-Effective Options for Reducing Fuel Use

A wide variety of options to reduce fuel consumption exist. However, they are of primary interest to the Air Force only if they offer cost-effective solutions. If implementing an approach costs more than the savings gained from fuel reductions, it does not make sense to implement it, in the absence of other benefits. In this chapter, we describe the options for reducing fuel use that are cost-effective purely from a fuel savings perspective. The options are presented in order of cost-effectiveness, from greatest to least. For each option, we describe the fuel saving approach, present our analysis of the approach, and then give the results of the analysis and our conclusions.

Engine-Out Taxiing

Aircraft can often taxi using fewer than all engines and thereby reduce their fuel burn without additional investment. This occurs because even the idle thrust of large turbofan engines is sufficient to move an aircraft at taxi speed on the ground.

Description

When surface conditions and aircraft weight are favorable, aircraft crew can delay starting all engines and taxi with fewer than all engines for fuel conservation. Engine-out taxiing can have additional benefits at low aircraft weight when the idle thrust of all engines far exceeds that required to taxi the aircraft and the crew may be required to apply braking force continually. Commercial carriers regularly use one-engine taxiing.[1] For example, JetBlue reports that it is able to conduct single-engine taxiing on 62 percent of its A320 flights systemwide.[2] Note that industry often refers to engine-out taxiing as single-engine taxiing, since many aircraft in its fleets are two-engine.

MAF aircraft operations procedures consistently direct delayed engine start; however, only the C-17 procedure suggests engine-out taxiing.[3] Our observations and conversations

---

[1] American Airlines Newsroom, undated; Delta Airlines, "Environmental Fact Sheet," online, updated January 2010; JetBlue, *2012 Responsibility Report*, undated (b).

[2] JetBlue, 2012.

[3] Air Force Instruction (AFI) 11-2C-130J, Volume 3, *C-130J Operations Procedures*, §14.5.14, December 8, 2009; AFI 11-2C-130, *C-130 Operations Procedures*, Vol. 3, §14.2.5, April 23, 2012; AFI 11-2C-17, Volume 3, *C-17 Operations Procedures*, §6.25.1, November 16, 2011; AFI 11-2C-5, *C-5 Operations Procedures*, Vol. 3, §14.2.3, February 24, 2012; AFI 11-2KC-10, Volume 3, *KC-10 Operations Procedures*, §14.2.3, August 30, 2011; AFI 11-2KC-135, Volume 3, *C/KC-135 Operations Procedures*, §14.4.2.5, September 18, 2008, certified current October 15, 2010.

with pilots show that MAF crews already do engine-out taxiing to some extent; however, its use is not universal.

Analytic Approach

Because most flight manuals do not report fuel consumption as a function of thrust during ground operations, our analysis assumes that there is a linear relationship between the fuel consumed and the number of engines operating. For example, we assume that two-engine taxiing consumes 50 percent less fuel than four-engine taxiing. In actuality, the fuel consumption of two-engine taxiing will be slightly higher than 50 percent because of the above-idle thrust requirements at certain points in the taxiing. Further, to recognize that not all sorties have conditions amenable to engine-out taxiing, we conservatively estimate that 50 percent of sorties could engine-out taxi. This is less than the 62 percent JetBlue is able to achieve with its A320 fleet, partially to account for the fact that some level of engine-out taxiing is already done in the MAF fleet. Also, this analysis assumes, on average, that aircraft taxiing is about 15 minutes to and 15 minutes from the end of runway (EOR) or 30 minutes per sortie, including most checklist activities. Finally, we always assume two-engine taxiing, although three-engine taxiing could be more appropriate in certain situations.

Although flight manuals provide information about total fuel consumption in ground operations for mission planning, they do not detail fuel consumption rates. As a result, our analysis assumed that the ground operations consumption was approximately 30 percent of the cruise fuel consumption at midpoint weight. This idealized ratio produces ground consumption rates that are approximately consistent with flight performance manual data and are consistent with the authors' experience operating a Boeing 747-400F simulator, although more complex estimations for taxi fuel flow exist.[4] Table 3.1 details the rates we used.

Results and Conclusions

Table 3.2 shows that engine-out taxiing can potentially save 8.1 MG of fuel, or $30.3 million at today's fuel price of $3.73 per gallon. Most of these savings come from the C-17 fleet as a result of its high annual sortie rates. Taxiing on fewer engines may require longer launch times to start all engines and allow adequate engine warm-up at the EOR. Additionally, crew launch procedures and order of checklist items may need to be revised to accommodate the procedure. This option places only a limited burden on maintenance or crew operations and is done extensively in the commercial sector. Language used in current

---

[4] Tasos Nikoleris, Gautam Gupta, and Matthew Kistler, "Detailed Estimation of Fuel Consumption and Emissions During Aircraft Taxi Operations at Dallas/Fort Worth International Airport," *Transportation Research Part D: Transport and Environment,* Vol. 16, No. 4, June 2011, pp. 302–308.

Table 3.1. Ground Operation Fuel Consumption

	Cruise Fuel Flow (ppm)	Taxi Fuel Flow (percent)	Taxi Fuel Flow (ppm)
C-130	66	30	20
C-17	265	30	80
C-5	349	30	105
KC-135	149	30	45
KC-10	301	30	90

Table 3.2. Cost-Effectiveness of Engine-Out Taxiing

	Neutral Additional Impacts			
	Annualized Fuel Savings (MG / Percent)	Annualized Engine-Out Taxi Cost ($FY13M)	Annualized Fuel Savings ($FY13M)	Break-Even Fuel Cost ($FY13M)
C-130	1.4 (1.6%)	Minimal	5.4	Minimal
C-17	4.3 (0.9%)	Minimal	16.2	Minimal
C-5	0.7 (0.7%)	Minimal	2.5	Minimal
KC-135	1.3 (0.7%)	Minimal	4.8	Minimal
KC-10	0.4 (0.4%)	Minimal	1.5	Minimal
Total	8.1 (0.8%)	Minimal	30.3	

operation procedures does accommodate engine-out taxiing or delayed engine start when conditions permit, but the instructions do not direct it. Incorporating engine-out taxiing into standard operating procedures can help institutionalize engine-out taxiing and save millions of gallons of fuel annually.

Optimum Flight Level and Speed

For any given aircraft flying at a particular weight, there is a flight level (FL) and speed that will maximize the specific range of that aircraft, minimizing fuel consumption per mile flown. As the aircraft consumes fuel, it becomes lighter and the fuel minimizing altitude and speed change. A well-known approach to saving fuel is continuous adjustment of the flight level and speed to operate at the most fuel efficient condition possible throughout the flight.

Description

In ideal weather conditions and absent constraints imposed by air traffic control, the most fuel efficient flying is achieved by doing a cruise climb. In cruise climb, the aircraft climbs to its best initial altitude and then continuously increases its altitude as it burns fuel throughout the flight. However, cruise climb is generally not available to aircraft, because flight levels are restricted by air traffic control agencies to discrete intervals, typically 2,000 feet apart or 4,000 feet at higher altitudes.[5] This restricts altitude changes to periodic step-climbs, resulting in long cruises at constant altitude. During these cruise periods, fuel use can be reduced by continuously adjusting speed (slowing down, generally) to achieve the best specific range for that altitude and instantaneous aircraft weight.

Recently, MAF has been directing its fleets to fly at fuel minimizing conditions.[6] This includes flying at the fuel-optimal speed, at least initially on reaching the cruise flight level. As will be seen, implementation of these directives and restrictions on cruise-climb flying results in modest fuel savings opportunities through FL and speed control.

Analysis Approach

In this analysis, we focused on the cargo fleet. Tanker aircraft have many more complex constraints on their operations and are not considered here. Using RAND-developed flight models, we modeled every GDSS mission in seven ways:

1. **Constant Level and Speed.** The aircraft climbs to its initial optimal altitude and remains there for the duration of the flight, maintaining the initial optimal speed throughout. This is the worst case for fuel efficiency that we examined.
2. **Constant Level, Best Speed.** The aircraft climbs to its initial optimal altitude and remains there but varies its speed throughout the flight to maximize fuel efficiency at that altitude.
3. **Single Climb, Constant Speed.** The aircraft makes a single 2,000-foot step-climb, roughly at mid-flight. At each FL, the aircraft initially adopts the optimal speed for that FL and maintains that speed until it changes altitude. Note that for a number of shorter flights, a step-climb is never preferred and for those flights, a constant FL is maintained throughout, as in scenario (1) above.
4. **Single Climb, Best Speed.** Same as (3), but the aircraft varies speed throughout the flight to maximize fuel efficiency.
5. **Optimal Climbs, Constant Speed.** Same as (3), but the aircraft makes a 2,000-foot step-climb whenever it is desirable to do so. Again, for some flights, the optimal number of step-climbs is zero.
6. **Optimal Climbs, Optimal Speed.** Same as (5), but with speed varying continuously throughout the flight.
7. **Cruise Climb.** The aircraft continuously increases its altitude throughout the flight so that its fuel efficiency is optimized at all times. This is the best case for fuel

---

[5] Elizabeth L. Ray, "Air Traffic Control," FAA Order JO7110.65U, Washington, D.C.: Federal Aviation Administration, December 16, 2011.

[6] See, e.g., AMC, *C-17 Cruise Speed Change*, Fuel Efficiency Bulletin 09-01, updated September 11, 2014.

efficiency and will represent the maximum theoretical savings that could be achieved.

In each case, we calculate the total fuel consumption over the year of flying, enabling us to identify the incremental savings of each improvement in procedures. Note that the calculations yield fuel used per mile flown. However, the flight times will vary slightly as well, so the fuel used per flight hour will not be exactly the same. For the range of realistic conditions of interest (conditions 3 through 6 in the list above), these flight time differences are very small and do not affect the analysis results.

It is important to recognize that these are idealized operations. In reality, the optimal FL and speed are a complex function of a large number of factors, including winds, weather, traffic, airport conditions, etc. Commercial carriers make these calculations and adjust their flight plans in real time. Such evaluations are beyond the scope of this report, and such calculations and real-time adjustments are beyond the capability of many military aircraft. The results here are intended to approximate the order of fuel savings that could be achieved by optimizing FL and speed, although additional savings are possible with real-time data.

Results

Figure 3.1 shows the total annual fuel consumption, relative to the fuel used when flying at constant FL and speed, for each of the seven flying conditions outlined above. The overall available savings can be significant. The spread between the worst case and the optimal cruise climb is less than 3 percent. Although the savings for individual flights can be much greater, a large number of short-distance flights do not benefit from in-flight climbs or speed adjustments, reducing the enterprise savings opportunity. Also, the tanker fleets are not included in these results primarily because their mission set often requires specified altitude or speed.

The realizable savings for the Air Force, absent real-time data, are likely smaller than 3 percent. The Air Force already directs its flight crews to fly in a fuel efficient manner. This includes climbing when favorable to do so. In discussions with current and former Air Force pilots, we found that at least a single 2,000-foot step is typical, when appropriate during a flight. This suggests that an appropriate baseline for current Air Force flying is at least the single step, constant speed condition.

At the other end of the scale, although cruise climb flying can bring significant fuel savings, it is generally not allowed by air traffic control. Therefore, the best condition that can be achieved for the vast majority of flights is 2,000-foot step-climbs when favorable

Figure 3.1. Fuel Consumption at Different Flight Level and Speed Combinations

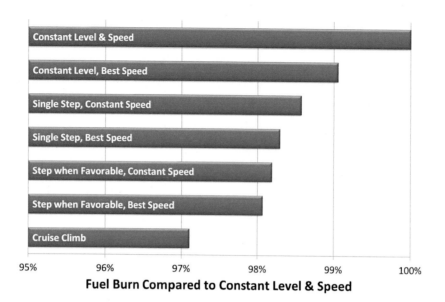

and continuously maintaining the best speed. The savings for this condition relative to the single-step, constant speed baseline is 0.51 percent, which likely represents the maximum potential savings available to the MAF cargo fleet.

Conclusions

Table 3.3 shows the fuel savings that aircraft could generate by performing 2,000-foot step-climbs whenever possible and continuously adjusting speed (step when favorable, best speed) as opposed to a single 2,000-foot step-climb with a constant speed (single-step, constant speed). We first recommend that the Air Force monitor and ensure that the use of a fuel minimizing flying condition is standard practice throughout its fleets. It should

Table 3.3. Cost-Effectiveness of Optimum Flight Level and Speed

	Neutral Additional Impacts			
	Annualized Fuel Savings (MG / Percent)	Annualized Optimum FL & Speed Cost ($FY13M)	Annualized Fuel Savings ($FY13M)	Break-Even Fuel Cost ($FY13M)
C-130	0.5 (0.5%)	Minimal	1.8	Minimal
C-17	2.4 (0.5%)	Minimal	8.9	Minimal
C-5	0.6 (0.7%)	Minimal	2.4	Minimal
KC-135	N/A	N/A	N/A	N/A
KC-10	N/A	N/A	N/A	N/A
Total	3.5 (0.4%)	Minimal	13.1	

further stress as standard procedure performing step-climbs whenever appropriate and adjusting speed. The Air Force may realize still greater savings by implementing real-time data collection and flight-optimization systems on MAF aircraft.

Basic Weight Reduction

The empty weight of an aircraft consists of the weight of the airframe, engines, propellers, rotors, and fixed equipment.[7] The basic weight of an aircraft is the empty weight of the aircraft plus unusable fuel, oil, oxygen, survival kits, and other equipment not consumed during flight.[8] As with any aircraft weight reduction, a reduction of aircraft basic weight will lead to increased fuel efficiency.

Description

Carrying additional weight entails a cost, as was shown in the cost-to-carry analysis. Any decrease in the total weight of an aircraft in flight will yield fuel savings based on the cost-to-carry analysis presented above. During the design phase of an aircraft, a great deal of attention is paid to reducing the aircraft basic weight.[9]

While aircraft of the same MDS have the same underlying design, the basic weight can vary from aircraft to aircraft. For example, the weight of an aircraft often increases with age because of foreign objects and dirt that collect and repairs and modifications.[10] There may also be design differences between aircraft, particular block changes, which account for weight differences. Identifying aircraft with higher-than-expected basic weight may allow for identification of weight reduction opportunities that could lead to fuel savings.

Commercial airlines are continuously looking for ways to reduce aircraft weight, including items beyond those listed in the basic weight. For example, Lufthansa recently stripped an Airbus A340 of almost four metric tons of items.[11] The Air Force has also looked for weight reduction savings.[12] As a specific example of this, the Air Force has begun to introduce electronic flight bags to reduce the amount of paper carried.[13] In

---

[7] Code of Federal Regulations, Title 14—Aeronautics and Space, Section 119.3 Definitions, January 1, 2003.

[8] Technical Order (T.O.) 1-1B-50, *Basic Technical Order for USAF Aircraft: Weight and Balance,* Tinker AFB, Okla.: 557 ACSS/GFEAC, April 1, 2008.

[9] Daniel P. Raymer, *Aircraft Design: A Conceptual Approach*, 4th ed., Reston, Va.: American Institute of Aeronautics and Astronautics, Inc., 2006.

[10] United States Department of Transportation Federal Aviation Administration Flight Standards Service, *Aircraft Weight and Balance Handbook*, Washington, D.C., 2007.

[11] Graham Warwick, "How Many Bin Bags to Empty an A340?" *Aviation Week*, blog post, March 20, 2013 (b).

[12] Amy McCullough, "Energy Effectiveness," *Air Force Magazine*, January 19, 2011.

[13] Matthew Stibbe, "U.S. Air Force Will Save $50M with iPad Electronic Flight Bags," *Forbes.com*, May 30, 2013.

addition, larger weight reduction efforts, such as removing excess galley equipment and redundant armor, are under way in the MAF fleet.[14] The Air Force has also made efforts to palletize certain aircraft equipment that is not needed for local sorties, loading the pallet only for sorties departing home station.

Analysis Approach

Analyzing the weight of each MAF aircraft in detail is beyond the scope of this work. A basic weight checklist record for a C-17 can contain over 100 items, and, we are not in a position to evaluate the necessity of carrying any one of these items, nor did we investigate the potential for reducing the weight of particular items. Such analysis would best be carried out by an entity such of AMC's FEO , and it has made significant progress in this regard.[15]

Our approach here is simply to develop a reasonable weight reduction estimate for the MAF fleet so that the value of reducing aircraft weight can be compared to the other options explored in the work. To do this, we compared the basic weight of individual aircraft as reported on Form B—Aircraft Weight Record to the MDS average.[16] We then determined the fuel savings that could be achieved if those aircraft with an above-average weight were brought down to the average weight.

Results

Figure 3.2 shows the deviations of individual aircraft basic weight from the average basic weight for the MDS. The aircraft are ordered in increasing basic weight. Surprisingly, the deviation does not increase significantly as the aircraft basic weight increases; rather, the deviation appears roughly constant for all MDS rather than proportional to the basic weight. The average deviation of basic weight from the average for the MDS is 1,258 pounds, and slightly more than half of the aircraft have a basic weight higher than the average for their MDS. If we reduced the basic weight of all aircraft heavier than the average for their MDS to the average weight, this would produce a net fuel savings roughly equivalent to reducing the weight of all aircraft by 655 pounds.

---

[14] DoE, 2012.

[15] Valerie Insinna and Yasmin Tadjdeh, "Air Force Making Headway on Fuel Efficiency Goals," *National Defense Magazine*, June 2013.

[16] T.O. 1-1B-50, 2008.

Figure 3.2. Basic Weight Deviation Between Aircraft of the Same MDS

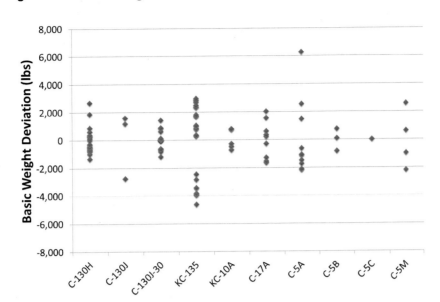

Conclusions

A basic weight reduction of 655 pounds for all aircraft in the MAF fleet would save 1.7 MG of fuel a year, which corresponds to an annualized savings of $6.2 million. As Table 3.4 shows, the savings are greatest in percentage terms for the C-130, since a 655-pound weight reduction represents a much larger proportion of the total aircraft weight than it does for the other aircraft. However, in terms of total fuel savings, the savings are greatest for the C-17 because of the fleet size and the relatively high amount of flying. We recommend that the Air Force continue to look for opportunities to reduce aircraft weight but recognize that extensive efforts to remove unnecessary equipment have already been undertaken and that only modest additional fuel savings are likely possible.

Table 3.4. Cost-Effectiveness of Basic Weight Reduction

	Neutral Additional Impacts			
	Annualized Fuel Savings (MG / Percent)	Annualized Lower Weight Cost ($FY13M)	Annualized Fuel Savings ($FY13M)	Break-Even Fuel Cost ($FY13M)
C-130	0.4 (0.4%)	Minimal	1.5	Minimal
C-17	0.6 (0.1%)	Minimal	2.3	Minimal
C-5	0.1 (0.1%)	Minimal	0.2	Minimal
KC-135	0.4 (0.2%)	Minimal	1.7	Minimal
KC-10	0.1 (0.1%)	Minimal	0.5	Minimal
Total	1.7 (0.2%)	Minimal	6.2	

Auxiliary Power Unit Use Reduction

Either aerospace ground equipment (AGE) or an aircraft's APU can provide electrical, pneumatic, or hydraulic power as well as air conditioning or heating for an aircraft. AGE is significantly more fuel efficient than APU; therefore, increasing the use of AGE in lieu of the aircraft's APU during ground operations can offer a substantial reduction in fuel consumption.

Description

AGE is gas- or electrical-powered equipment that generally performs a single function. For example, during a typical engine start, a gas-operated power unit can provide electrical power, a gas turbine compressor can provide pressurized air to rotate aircraft engines during start, and an air-conditioning unit or heater unit can provide cooling or heating, as needed. An aircraft APU is a gas turbine engine that can provide the same functions as these AGE but its fuel consumption is generally more than three times that of a ground electrical power unit and an air-conditioning unit, combined.[17]

Commercial airlines have been aggressive in their attempt to reduce APU use. United Airlines notes that ground power and cooling systems can be up to eight times more efficient than the APU.[18] American Airlines notes that using the APU ground power and cooling, it is able to save more than 6 MG of fuel a year.[19] FedEx quantifies its savings on a per-flight basis, and finds that it can reduce APU use by 1.5 hours by using ground power.[20] Of course, the Air Force has different objectives and constraints than commercial carriers, and therefore applying industry savings directly is not possible.

In 2009 and again in 2010, AMC mandated both aircrew and maintenance personnel maximally use AGE in a manner that was consistent with technical order guidance and mission requirements.[21] Additionally, each aircraft's operating procedure directs minimizing the use of APU when possible.[22] Despite this guidance, 2011 APU usage data indicates significant potential for additional savings on some MAF aircraft.[23]

---

[17] AMC, Fuel Efficiency Office, no title, Bulletin 09-02, updated September 11, 2014.

[18] United Airlines, "United Eco-Skies: A Commitment to the Environment," online, December 11, 2011.

[19] American Airlines Newsroom, undated.

[20] FedEx Express, "FedEx and the Environment," fact sheet, online, June 2013.

[21] AMC/A4M released a message in January 2009 and a logistics interest item effective November 2010, and AMC/A3 released a message and two flight crew information files (January 2009 and June 2010), all similarly directing the use of AGE and restricting APU use to situations when absolutely necessary.

[22] Air Force Instruction 11-2C-17, November 16, 2011; AFI 11-2C-5, February 24, 2012; AFI 11-2C-130J, December 8, 2009; AFI 11-2KC-10, August 30, 2011; AFI 11-2KC-135, September 18, 2008.

[23] The 2011 APU usage data were extracted from AMC, "Reduce APU Use," *Reduce APU Use Program Evaluation,* Scott AFB, Ill.: AMC/A4M, March 2012.

Analytic Approach

According to data provided by the Fuel Efficiency Office, Table 3.5 details the relative consumption of various APU and AGE.[24] This assumed that a ground power unit and either an air-conditioning unit or heater unit were required in lieu of an APU, even though both may not always be required.[25] For the C-5, which has two APUs, we assumed that the aircraft would require two power units and two air-conditioning units.

To estimate the potential for reducing APU use, we first corrected for temperature effects. Greater APU use is generally seen in unusually hot or cold weather. We identified the major operating location that, for each MDS, had the lowest average APU use.[26] Although this may seem optimistic, since one would assume that average APU use would be significantly lower at major bases than at austere locations, the data do not support this assumption. With this approach, we found no significant savings for the C-130 or KC-135. For the C-17, we found that the APU use at the ten most frequented airfields is only 3.5 minutes less than the overall average. However, APU use for C-17 flights from Joint Base Charleston is almost 30 minutes shorter than the overall average. If the APU practices established at Joint Base Charleston could be replicated elsewhere, fuel saving would improve significantly. Similarly, we might expect austere locations, such as those in Afghanistan, to experience higher APU use than stateside locations, but again the data do

Table 3.5. APU and AGE Fuel Consumption

	Fuel Consumption (gph)
C-5 APU (2 APUs per aircraft)	44.4 each
C-17 APU	59.2
KC-10 APU	59.2
C-130 APU	51.7
KC-135 APU	37.0
Air Cart (A/M32A-95 GTC)	15.8
Ground Power Unit (B809D)	6.0
Air Conditioning Unit	7.3

---

[24] AMC, Fuel Efficiency Office Bulletin 09-02.

[25] This is the same assumption made to estimate savings in AMC, 2012.

[26] We define major operating locations as the ten sites with the greatest frequency of aircraft operations for a given MDS.

not support this. For example, average C-17 APU use in Afghanistan is four minutes less than the overall average. From this, there does not appear to be a fundamental reason that overall APU use cannot be reduced to that of Joint Base Charleston, the identified target location.

The use of the APU for maintenance actions was not considered as part of this analysis. However, reducing maintenance use of the APU could be a cost-effective fuel reduction option.

Results and Conclusions

Table 3.6 indicates the overall average APU use for each MDS, the major operating location with lowest average APU use, and the potential savings garnered from reducing the overall average to match the lowest-use location.[27]

Figure 3.3 plots the average APU use for the C-5, C-17, and KC-10 and the target use based on the operating location with the lowest average APU use. The fuel and cost savings associated with reaching the target APU use per sortie are given for each MDS. For reference, the horizontal black bar indicates the representative commercial sector practice and suggests potential for further savings beyond the best practices at Air Force operating locations.

Reducing average APU at all airfields to that achieved at the target locations could save the Air Force more than 1 MG of fuel per year. As shown in Table 3.7, this equates to $4.7 million of fuel cost savings at today's fuel price of $3.73 per gallon with most of these savings coming from the C-17 fleet because of its high annual sortie rates.[28] Discussions with operators indicate that there are sufficient AGE units available to meet an additional

Table 3.6. Average and Lowest APU Use Data

	Average APU Use per Sorite (mins)	Major Location with Lowest Use	Potential Reduction per Sortie (mins)
C-17	110.8	Joint Base Charleston	28.6
C-5	116.3	Stewart Air Force Base	24.4
KC-10	75.8	Travis Air Force Base	9.6

SOURCE: AMC, 2012.

---

[27] AMC, 2012.

[28] This savings equates purely to fuel savings. It does not include secondary effects, such as an increase in maintenance cost for ground equipment and decrease in maintenance costs for APUs.

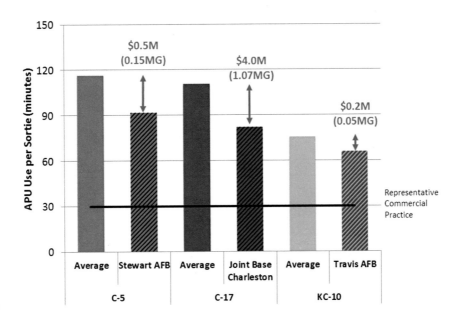

Figure 3.3. APU Usage Rates Compared to APU Usage at Lowest Operating Location

Table 3.7. Cost-Effectiveness of APU Usage Reduction

	Neutral Additional Impacts			
	Annualized Fuel Savings (MG / Percent)	Annualized Reduced APU Use Cost ($FY13M)	Annualized Fuel Savings ($FY13M)	Break-Even Fuel Cost ($FY13M)
C-130	N/A	N/A	N/A	N/A
C-17	1.1 (0.2%)	Minimal	4.0	Minimal
C-5	0.1 (0.2%)	Minimal	0.5	Minimal
KC-135	N/A	N/A	N/A	N/A
KC-10	0.0 (0.0%)	Minimal	0.2	Minimal
Total	1.3 (0.1%)	Minimal	4.7	

demand; therefore, an investment in additional units would not necessarily be required to begin realizing these savings. As AGE use increases, there might be a commensurate increase in demand for maintenance manpower to ensure that ground units are readily available to service aircraft and this would need to be explored further. To realize these savings, maximal use of AGE should be embedded into training and integrated into standard flightline procedures to help develop a culture of fuel savings. Detailed analysis should be conducted to further determine how APU use can approach that of the target bases. Also, intensified monitoring and auditing, such as quality assurance and compliance inspections, could help the Air Force fuel conservation efforts.

Load-Balancing Improvement

For aircraft to be stable in flight, the center of gravity (CG) has to be forward of the center of lift.[29] However, this positioning creates inefficiency because of the downward lift on the horizontal tail. As the aircraft CG is moved farther forward, more downward lift is required on the tail, decreasing the aerodynamic performance. Because of this, maintaining proper CG location is important for maximizing fuel efficiency.

Description

The CG of an aircraft can be varied during flight by moving fuel forward or aft.[30] Movement of the CG within the allowable range can increase or decrease specific range by about 2 percent.[31] This is because as the CG is moved farther forward more trim— downward force on the horizontal tail—is required to keep the aircraft in level flight.[32] It is also possible to control the CG location by shifting loads within the cargo compartment. The Air Force has sought ways to improve the CG location, including by automation of Form F.[33]

Analysis Approach

To estimate the extent of possible CG improvement and the resulting fuel savings, we derived a detailed estimate for the C-17 and applied these results to other aircraft. Fuel tracker data maintained by the AMC FEO include CG location. We analyzed these data to determine what percentage of sorties had a significant payload and had a CG forward of the nominal location. For these sorties, we calculated the average CG location. We then calculated the improved specific range assuming that all of these sorties had instead been flown with a nominal CG. The CG location has no effect on fuel burned during ground operations. We estimated the improvement in specific range to be about 0.3 percent for every 1 percent change in mean aerodynamic chord.[34]

---

[29] Daniel P. Raymer, *Aircraft Design: A Conceptual Approach*, 3rd ed., Reston, Va.: American Institute of Aeronautics and Astronautics, 1999.

[30] Airbus, "Getting to Grips with Fuel Economy," *Flight Operations Support & Line Assistance, Airbus Customer Services*, No. 3, July 2004.

[31] Airbus, 2004; MAC, 1976.

[32] MAC, 1976.

[33] Harold Smoot, "AWBS: Automated Weight & Balance System," Lockheed Martin Corporation, undated. Also, see United States Department of Defense, Executive Services Directorate, "Weight and Balance Clearance Form F—Transport," August 1996.

[34] The mean aerodynamic chord is the distance between leading and trailing edges of the wing, accounting for both the wing's sweep and taper.

Results

We analyzed one year of data and found that 35 percent of C-17 sorties had a load greater than or equal to 5,000 pounds. Obviously, flights with little or no cargo are not candidates for improved load balancing, since there is no load to balance. Of those sorties with a load greater than or equal to 5,000 pounds, 74 percent had a CG forward of the nominal CG location of 37 percent of the mean aerodynamic chord, with the average at 35.68 percent mean aerodynamic chord. This corresponds to an average improvement of specific range of just over 0.1 percent. After determining the improvement achievable for the C-17, we applied similar factors to the other aircraft in the MAF fleet, as shown in Table 3.8. The same improvement in specific range results in a different fuel reduction for the different MDS when modeled across the entire spectrum of flight profiles. For example, the C-130 has the lowest percentage fuel reduction because of its relatively short sortie distances, meaning that ground operations and climb contribute to a greater proportion of the sortie fuel burn.

Conclusions

Given that average CG locations are not that far from the nominal location and that 65 percent of sorties fly near empty, the potential fuel savings is less than 1 MG per year, as is shown in Table 3.9. Some of this savings may not be realizable because of loading limitations, intermediate stops, and cargo uncertainty. However, we recommend that the Air Force continue to improve CG location through collecting better data and using load-balancing software.

Table 3.8. Fuel Savings from Improved CG Location

	Average CG Improvement (Percent MAC)	Specific Range Improvement (Percent)	Fuel Reduction (Percent)
C-130	0.34	0.11	0.06
C-17	0.34	0.11	0.09
C-5	0.34	0.11	0.10
KC-135	0.34	0.11	0.09
KC-10	0.34	0.11	0.09

Table 3.9. Cost-Effectiveness of Load-Balancing Improvement

	Neutral Additional Impacts			
	Annualized Fuel Savings (MG / Percent)	Annualized Load Balancing Cost ($FY13M)	Annualized Fuel Savings ($FY13M)	Break-Even Fuel Cost ($FY13M)
C-130	0.1 (0.1%)	Minimal	0.2	Minimal
C-17	0.5 (0.1%)	Minimal	1.8	Minimal
C-5	0.1 (0.1%)	Minimal	0.3	Minimal
KC-135	0.2 (0.1%)	Minimal	0.6	Minimal
KC-10	0.1 (0.1%)	Minimal	0.4	Minimal
Total	0.9 (0.1%)	Minimal	3.3	

Technical Stop Addition

A technical stop is essentially a refueling stop where no cargo is loaded or offloaded.[35] Refueling along a route, even though the aircraft is capable of flying the route nonstop had it loaded more fuel initially, can reduce the total amount of fuel burned because the aircraft has a lower average fuel weight.

Description

To understand the benefit of adding a technical stop, consider only the cruise portion of flight. The distance an aircraft can travel can be found by integrating the aircraft's specific range—the distance an aircraft can travel on a unit of fuel. Because of the relationship between increased fuel burn and increased fuel weight, we find that considering only the cruise segment, flying two half segments consumes less total fuel than flying one full segment.[36] This simple analysis includes only cruise. If we include fixed fuel consumption, such as start, taxiing and take-off, climb, and descent, this relationship will no longer hold for short-distance flights. Specifically, for the technical stop addition to be favorable, the cruise fuel savings of flying two legs will need to be greater than the additional fuel burned by executing an additional landing, stop, and takeoff.

---

[35] ICAO, *ICAO DATA+Glossary*, online, undated.

[36] The specific range is inversely proportional to the aircraft weight, $\frac{dR}{dW} \propto \frac{1}{W}$.[36] Integrating the Breguet range equation gives us that $R \propto \ln \frac{W_L + F}{W_L}$, where W_L is the landing weight and F is the weight of the fuel burned. This tells us that the amount of fuel burned in cruise can be estimated by $F = W_L \left(e^{R/\psi} - 1 \right)$, where ψ is the constant of proportionality and involves characteristics of the aircraft. It is important to note that the range appears in the exponential. We can now compare two cases: In the first we fly a distance, D, while in the second we fly half that distance twice. That is to say, $\frac{F_2}{F_1} = \frac{2\left(e^{D/2\psi} - 1\right)}{e^{D/\psi} - 1} = \frac{2}{1 + e^{D/2\psi}}$.

Analysis Approach

Using RAND-developed flight models, we modeled every GDSS mission in two ways. The first was to model them as they were actually flown; the second was to model twice the number of missions but with every mission covering half the distance. We then compared these two values for each mission and determined the minimum fuel burn.

Results

The RAND flight model results show that all aircraft can sometimes benefit from an additional technical stop, which we found to be roughly independent of payload and primarily based on distance. Obviously, for short-distance flights, adding a technical stop has a large fuel penalty associated with additional ground time and climb. Short-distance flights also see the least improvement in cruise efficiency because the fuel ratio derived earlier has the distance in an exponential in the denominator. Figure 3.4 shows the ratio of total fuel for the case with an additional technical stop to the case with no additional technical stop, which we term the fuel burn fraction. A technical stop reduces fuel use for missions where the fuel burn fraction is less than 100 percent. As the chart shows, the C-130H drops below this threshold only slightly before its maximum range. In fact, only 1.4 percent of C-130H sorties would benefit from an additional technical stop, and no C-130J sorties would benefit. However, 12 percent of C-17 sorties and 23 percent of C-5 sorties could benefit. Further, any additional aircraft stops will require ground launch and recovery

Figure 3.4. Technical Stop Fuel Burn Fraction Versus Total Flight Distance

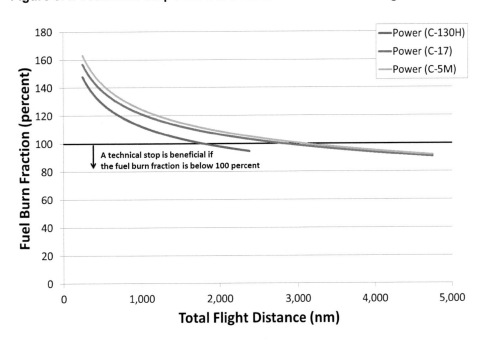

maintenance actions, in addition to refueling the aircraft. Further, stops that are not at established en route locations could require more logistics than that required at an established en route location.

Conclusions

Using the most optimistic assumption of adding a technical stop to every sortie that would benefit, no matter how small the benefit, we find savings for the C-17 and C-5 and a negligible savings for the C-130. The potential for over 8 MG of fuel savings is highlighted in Table 3.10. However, we note that the negative consequences are significant. Performing an additional technical stop increases the demand for personnel, resources, and time and could cause issues with crew duty day. Additional landings also stress the aircraft and increase the likelihood of a mechanical failure. Furthermore, this analysis assumes that an airfield is available near the midpoint and no additional flying would need to be done to reach the technical stop location. We recommend that the Air Force consider additional technical stops when practical on the longest sorties, perhaps those longer than 4,500 nautical miles (nm), and where an en route refueling location is already established, but first perform a full cost analysis to quantity the additional cost and risk associated with additional technical stops.

Table 3.10. Cost-Effectiveness of Additional Technical Stop

	Annualized Fuel Savings (MG / Percent)	Annualized Ground Towing Cost ($FY13M)	Annualized Fuel Savings ($FY13M)	Break-Even Fuel Cost ($FY13M)
			Negative Additional Impacts	
C-130	0.0 (0.0%)	Minimal	0.0	Minimal
C-17	6.7 (1.3%)	Minimal	24.9	Minimal
C-5	1.6 (1.6%)	Minimal	5.8	Minimal
KC-135	N/A	N/A	N/A	N/A
KC-10	N/A	N/A	N/A	N/A
Total	8.2 (0.8%)	Minimal	30.8	

Continuous Descents

On most approaches, aircraft descend in a stepped-approach fashion. This means that aircraft spend more time flying at lower altitudes than is necessary based solely on aerodynamics. This additional time at low altitudes translates to additional fuel burn.

Description

A continuous descent operation (CDO) or continuous descent approach (CDA) is one in which an arriving aircraft descends continuously, to the greatest possible extent, by employing minimum engine thrust before the final approach point.[37] A CDA has two major advantages over conventional stepped descents: reduced fuel consumption and reduced noise. In a CDA, the aircraft flies at a higher altitude (ideally but not necessarily its preferred cruise altitude) for longer, whereas in a conventional approach, the aircraft descends in one or more steps and must cruise at low altitude for significant distances. The difference in descent profiles is shown schematically in Figure 3.5

In practice, CDAs are not generally available, because of air traffic control limitations and procedures. In addition, there are also hardware and software requirements such as incorporation of wind data in the flight computer. Recently, the FAA developed several initiatives to implement optimized profile descents (OPDs), a way to allow maximum practical use of a CDA.[38] As of 2010, Los Angeles International Airport (LAX) had become the airport with the first published, publicly available OPD procedure in the United States, and approximately 40 to 50 percent of the approaches at LAX were by OPD. The FAA estimates fuel savings of 25 gallons per flight that uses an OPD.

Figure 3.5. Comparison of Continuous and Conventional Descent Approaches

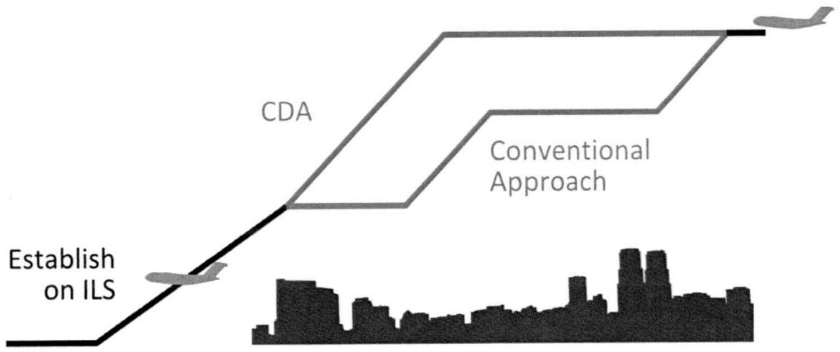

SOURCE: Adapted from the European Organisation for the Safety of Air Navigation.

Analysis Approach

In this analysis, we modeled every GDSS mission with a simplified stepped descent and flew it again with a CDA. For the stepped descent, we used a single 15-nm long step at 5,000 feet, based on a common step used in approaches into Hartsfield-Jackson Atlanta International Airport.[39] We did not include additional steps (such as a step at 12,000 feet) on the assumption that even in places where CDAs are implemented, they are unlikely to

---

[37] ICAO, *Continuous Descent Operations (CDO) Manual,* Doc. 9931, 1st ed., 2010 (a).

[38] Federal Aviation Administration, "Atlantic Interoperability Initiative to Reduce Emissions (AIRE)," 2010 AIRE Workshop, Brussels, Belgium, 2010.

[39] Ian Wilson and Florian Hafner, "Benefit Assessment of Using Continuous Descent Approaches at Atlanta," paper presented at 24th Digital Avionics Systems Conference, Washington, D.C., October 30, 2005.

be accomplished all the way from cruising altitude but rather from a lower holding altitude particular to each airport.[40] We calculated the fuel savings achieved via CDA for each MAF aircraft.

As an alternative approach, we calculated the fuel savings according to the FAA's mixed-fleet estimate of 25 gallons per flight. The MAF aircraft mix is undoubtedly very different from that analyzed by the FAA, but the comparison is useful as a check on the order of magnitude of achievable savings.

Results

Results of both analyses are shown in Table 3.11. Note that the FAA estimate results in only about 55 percent of the fuel savings calculated through the stepped-descent analysis and gives quite different results for a number of platforms, particularly the tanker fleets. For our final results, shown in Table 3.11, we use our stepped-descent analysis but present the simpler estimate as a point of comparison. "Fuel Savings with Stepped-Descent Analysis" is obtained by running every 2012 GDSS flight with CDA and with stepped descent. The "Fuel Savings with FAA Estimate" applies the FAA's fuel savings of 25 gallons per flight to 2012 GDSS flying.

Table 3.11. Comparison of Two Approaches to Calculating Fuel Savings Achieved by CDA

	Fuel Savings with Stepped-Descent Analysis (percent)	Fuel Savings with FAA Estimate of 25 gal/flight (percent)
C-130	1.10	2.67
C-17	0.53	0.34
C-5	0.32	0.19
KC-135	1.43	0.65
KC-10	0.23	0.48

Conclusions

In general, CDA is not something that the Air Force can implement unilaterally. Since implementation of CDA requires changes to air space and the approval of other agencies, we assess it as having negative additional impacts for the Air Force. The ability to use OPDs at most commercial airports will be determined by air traffic control procedures and the prevailing local traffic and weather conditions. To the extent that the Air Force can

---

[40] See, for example, United Kingdom Civil Aviation Authority, "Basic Principles of the Continuous Descent Approach (CDA) for the Non-Aviation Community," London: Environmental Research and Consultancy Department (ERCD) of the Civil Aviation Authority, undated.

prepare its aircraft and pilots for CDA procedures, it may be able to implement CDA at military bases and other airfields under its control, the achievable fuel savings will very likely justify this implementation. A full implementation of CDA, although unlikely, could result in nearly 12 MG of fuel savings a year, as is seen in Table 3.12. However, use of CDA would require additional aircraft capabilities, which are not quantified here.

Table 3.12. Cost-Effectiveness of Continuous Descents

	Annualized Fuel Savings (MG / Percent)	Annualized Optimum Descent Cost ($FY13M)	Annualized Fuel Savings ($FY13M)	Break-Even Fuel Cost ($FY13M)
			Negative Additional Impacts	
C-130	1.0 (1.1%)	Minimal	3.8	Minimal
C-17	2.7 (0.5%)	Minimal	10.0	Minimal
C-5	0.3 (0.3%)	Minimal	1.1	Minimal
KC-135	2.7 (1.4%)	Minimal	10.1	Minimal
KC-10	0.3 (0.2%)	Minimal	1.0	Minimal
Total	7.0 (0.7%)	Minimal	25.9	

Vortex Surfing

Vortex surfing is a form of formation flying designed to reduce fuel use in which an aircraft flies in the vortices generated by a leading aircraft. These vortices are often referred to as lift-induced vortices because they result from the pressure differences between and along the top and bottom surfaces of the wing that are responsible for the generation of lift.

Description

Figure 3.6 shows two vortices generated from a leading aircraft and the upwash and downwash generated by these vortices. The figure shows three key areas: the downwash area between the two vortex cores, the vortex cores themselves, and the upwash area outboard of the vortex cores. Aircraft flying in the downwash region would experience a loss of altitude or a decrease in the rate of climb. In the cores themselves, aircraft would experience a significant imposed roll. Outboard of the vortex cores, aircraft would experience an upwash generated by the vortices; flying in this area would reduce the amount of thrust required to maintain level flight, thereby producing fuel savings.

In general, aircraft should avoid flying into the vortex, or wake, of another aircraft because of the associated large changes in flow direction and speed. As Figure 3.6 demonstrates, an aircraft flying across the vortices would experience large structural load factors. Rules regarding aircraft separation direct pilots who believe they may be flying in the vortex of another aircraft to move away from it. Incidents associated with aircraft flying

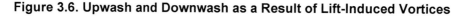

Figure 3.6. Upwash and Downwash as a Result of Lift-Induced Vortices

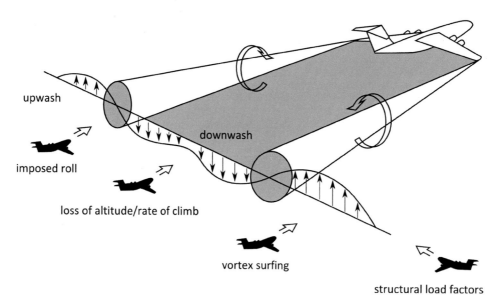

SOURCE: Adapted from William A. McGowan, "Aircraft Wake Turbulence Avoidance," paper presented at 12th Anglo-American Aeronautical Conference, Calgary, July 7–9, 1971.

in the wake of another aircraft are not uncommon. For example, American Airlines Flight 587 crashed when the first officer incorrectly responded to wake turbulence.[41]

In 2001, NASA demonstrated that an F/A-18 could fly in the vortex of another F/A-18 and experience more than an 18 percent reduction in fuel flow.[42] More recently, in flight tests conducted in 2011, 2012, and 2013, two programs known as Cargo Aircraft Precision Formations for Increased Range and Efficiency and Surfing Aircraft Vortices for Energy showed cruise fuel flow reductions upwards of 10 percent.[43] Initial indications are that the ride is rougher, but acceptable, and damaging effects on the airframe have not been observed.[44]

---

[41] Allied Pilots Association, Submission of the Allied Pilots Association to the National Transportation Safety Board Regarding the Accident of American Airlines Flight 587 at Belle Harbor, New York, November 12, 2001, NTSB DCA02MA001, undated.

[42] Ronald J. Ray, Brent R. Cobleigh, M. Jake Vachon, and Clinton St. John, "Flight Test Techniques Used to Evaluate Performance Benefits During Formation Flight," Edwards AFB, Calif.: National Aerospace Laboratory, Dryden Flight Research Center, NASA/TP-2002-210730, August 2012.

[43] Joe Pahle, Dave Berger, Michael W. Venti, James J. Faber, Chris Duggan, and Kyle Cardinal,, "A Preliminary Flight Investigation of Formation Flight for Drag Reduction on the C-17 Aircraft," briefing, Washington, D.C.: National Aeronautics and Space Administration, NASA 20120007201, March 7, 2012; Roger Drinnon, "'Vortex Surfing' Could Be Revolutionary," *Air Force Print News Today*, October 10, 2012; Paul D. Shinkman, 'Vortex Surfing' Could Save Military Millions," *U.S. News and World Report*, August 13, 2013; and Joe Pappalardo, "Vortex Surfing: Formation Flying Could Save the Air Force Millions on Fuel," *Popular Mechanics*, July 17, 2013.

[44] Pahle et al., 2012.

Analytic Approach

Flights tests have indicated the potential for reducing cruise fuel flow for a trailing C-17 by up to 10 percent. For the purposes of this analysis, we will conservatively assume that vortex surfing nets an 8 percent reduction in fuel flow throughout cruise. Fuel flow reductions of 10 or more percent seem to only occur in a limited "sweet spot" region, which can be difficult to maintain.[45] Further, there are time periods at the beginning and end of cruise where the aircraft is moving into and out of position and no fuel savings can be realized, and in fact some additional fuel burn may be required for maneuvering the aircraft into the correct relative position. Although fuel flow reductions have been demonstrated only on the C-17, we assume that both C-5 and C-130 aircraft could achieve similar reductions. We did not include tankers in this analysis because of the refueling activities they may be executing en route.

Two important factors limit the benefits of vortex surfing and need to be accounted for. First, to take advantage of vortex surfing, two aircraft must be flying the same route at the same time. We make a few key assumptions in this regard. First, we assume that aircraft must be flying the same city-pair on the same day. Second, we assume that the aircraft must be of the same MDS. Third, we assume that aircraft cannot trail each other in series.[46] This means that we assume that at most there could be two staggered aircraft following a single lead aircraft.[47] There exists potential to expand beyond these assumed limitations, of course. Sorties could be combined over multiple days by delaying or moving forward some flights. Also, aircraft do not necessarily need to be flying the same city-pair to take advantage of vortex surfing. For example, two aircraft heading for Hawaii could depart from different continental U.S. bases and then meet over the ocean and take advantage of vortex surfing for a significant portion of the flight. Finally, the aircraft do not need to be the same MDS. In fact, NASA demonstrated the potential for vortex surfing with two drastically different aircraft, an F/A-18 and a DC-8.[48]

While the assumption that the aircraft must fly the same city-pair on the same day may seem conservative, it is important to recognize that for operational reasons even this may prove difficult. For example, there may be time-sensitive cargo such that sorties cannot be shifted to allow both to depart nearly simultaneously, or there may be maintenance issues that delay one aircraft. In addition, vortex surfing with aircraft of two or more MDS would require an additional level of software and communications capability.

---

[45] Pahle et al., 2012.

[46] Gordon Lubold, "How Geese Will Save the Air Force Millions of Dollars," *ForeignPolicy.com*, July 18, 2013.

[47] NASA, "Sky Surfing for Fuel Economy," online, July 24, 2003.

[48] NASA, 2003.

The second issue is the fact that cruise only accounts for a portion of the total fuel burn. In addition to cruise, where vortex surfing can reduce fuel use, there is also fuel required for aircraft start, taxi, takeoff, climb, descent, and shutdown.

Results

We were able to identify how frequently multiple aircraft of the same MDS flew the same city-pair on the same day in FY 2012, and what proportion of fuel was burned in cruise for those flights. We found that vortex surfing was an option for 32 percent of sorties. Of these, 61 percent had two aircraft flying the same city-pair on the same day. In the case of two aircraft flying the same city-pair, only one can be in the trailing position and experience fuel savings. In addition, we find that the average distance of sorties where two or more aircraft are flying the same city-pair on the same day is 1,181 nm, which is shorter than the overall average C-17 sortie of 1,359 nm.

For the C-17, 15 percent of the total miles flown can be flown while trailing another aircraft. Applying an 8 percent cruise flow reduction to these sorties nets a 5.3 percent total fuel burn reduction for the trailing aircraft. Across the entire C-17 fleet, this translates to a reduction of 0.8 percent in fuel use.

We can similarly compute these figures for the C-130 and the C-5, the results of which are shown in Table 3.13. The C-130 has a significantly lower per-sortie fuel reduction than the other aircraft because the C-130 average sortie distance for city-pairs with two or more aircraft flying on the same day is only 256 nm. This means that a relatively smaller portion of fuel is being consumed in cruise where vortex surfing can net savings. Conversely, the C-130 has the greatest proportion of flights that can take advantage of vortex surfing. Overall, the total fuel reduction potential from vortex surfing varies between 0.5 and 0.8 percent, depending on MDS.

Table 3.13. Vortex Surfing Saving Calculations for C-130, C-17, and C-5

	Cruise Fuel Reduction	Sortie Fuel Reduction	Miles Flown in Trailing Position	Net Fuel Reduction
C-130	8.0%	2.6%	19%	0.5%
C-17	8.0%	5.3%	15%	0.8%
C-5	8.0%	6.4%	10%	0.6%

Conclusions

Flight tests have demonstrated that a C-17 trailing another C-17 can experience as much as a 10 percent reduction in cruise fuel flow.[49] For the C-17, the ability to vortex surf would require only a software update to the formation flight system.[50] Such an update could presumably be done as part of a larger software update and would likely not cause any significant additional development costs. One key area of concern is any possible structural fatigue that vortex surfing could cause on the aircraft. Earlier indications are that there are no airframe or engine life cycle issues.[51] However, given the reports of varying ride quality,[52] we strongly recommend that this be explored in great detail. As past work shows,[53] even a small reduction in aircraft service life can have costs significantly greater than any potential fuel savings.

As Table 3.14 shows, the savings from vortex surfing could yield 5.1 MG of fuel a year, with 4.1 MG of that coming from the C-17 fleet alone. At today's fuel price of $3.73 per gallon, this would be slightly under $20 million of fuel savings a year. We have flagged this option as having negative additional effects because of the logistics of coordinating flights, additional training that would be required by crews, potential safety issues, and potential aircraft structural issues. However, given the savings potential and the amount of research that has already been done, we recommend continuing research into the area of vortex surfing and how to best implement it in the MAF fleet, presumably first with C-17s. Additional data and research might reveal ways to negate potential negative effects.

Table 3.14. Cost-Effectiveness of Vortex Surfing

	Annualized Fuel Savings (MG / Percent)	Annualized Vortex Surfing Cost ($FY13M)	Annualized Fuel Savings ($FY13M)	Break-Even Fuel Cost ($FY13M)
		Negative Additional Impacts		
C-130	0.4 (0.5%)	Minimal	1.7	Minimal
C-17	4.1 (0.8%)	Minimal	15.2	Minimal
C-5	0.6 (0.6%)	Minimal	2.2	Minimal
KC-135	N/A	N/A	N/A	N/A
KC-10	N/A	N/A	N/A	N/A
Total	5.1 (0.5%)	Minimal	19.1	

---

[49] Pahle et al., 2012.

[50] Drinnon, 2012.

[51] Graham Warwick, "C-17s Go Surfing, to Save Fuel," *aviationweek.com* blog post, October 12, 2012.

[52] Drinnon, 2012.

[53] Mouton, 2013.

Paint Weight Reduction

Painting of aircraft is a primary means of preventing aircraft corrosion.[54] However, paint adds weight to the aircraft.[55] If the weight of paint on the aircraft could be reduced without jeopardizing the corrosion protection, there would be some potential for fuel savings.

Description

The Air Force primarily paints its aircraft for corrosion prevention, but paint also improves aircraft appearance.[56] As with any additional weight on the aircraft, there is a fuel burn penalty for using paint, which can be understood by looking at the cost-to-carry for the particular aircraft. American Airlines' aircraft historically featured a largely unpainted, polished surface, presumably for weight reduction. However, American Airlines' newest livery has moved away from a polished aircraft to a painted aircraft.[57] This suggests that at today's fuel prices, the weight savings of not painting aircraft do not offset the additional costs of maintaining polished surfaces.

Painting of aircraft involves two primary components, the primer and the top coat, both of which increase the aircraft's weight.[58] Even current composites require painting to prevent water damage.[59]

Painting is primarily done for corrosion protection. The Air Force spends over $1 billion annually on corrosion protection.[60] This indicates that even the smallest loss of corrosion protection would easily outweigh the savings associated with a modest reduction in aircraft weight. Therefore, any attempt to reduce aircraft paint weight must not compromise the corrosion resistance of the aircraft.

Analysis Approach

However, it may be possible to achieve weight savings through the development of new paint processes, including primer-free paint processes. We hypothesize that such improvements could reduce the total paint weight by 50 percent. Naturally, our analysis

---

[54] Naval Air Systems Command, *Cleaning and Corrosion Control,* Volume I: *Corrosion Program and Corrosion Theory*, Patuxent River, Md., NAVAIR 01-1A-509-1, March 1, 2005.

[55] Dan Hansen, "Painting Versus Polishing of Airplane Exterior Surfaces," *AERO Magazine*, online, undated.

[56] United States Department of Defense (DoD), *Air Force Aircraft Painting and Corrosion Control,* Washington, D.C., Inspector General Audit Report No. 96-062, January 1996.

[57] Andrew Bender, "American Airlines' Makeover: Design Pros Weigh In," *Forbes.com*, January 21, 2013.

[58] Gary N. Carlton, "Aircraft Corrosion Control: Assessment and Reduction of Chromate Exposures," Brooks Air Force Base, Tex.: Air Force Institute for Environment Safety, June 2000.

[59] Jim Rowbotham, "Coatings for Composites," *AERO Magazine*, February 13, 2012.

[60] DoD, 1996; Wayne Crenshaw, "Corrosion Office Helps Prolong Life of Aircraft," Robins Air Force Base, Ga.: 78th Air Base Wing Public Affairs, February 10, 2009.

assumes that a primer-free paint process would continue to protect the airframe from corrosion.

To compute the total weight we used published estimates for the painted area[61] and estimated the weight of paint and primer to be about 0.025 pounds per square foot.[62] Estimates for the paint and primer weights are given in Table 3.15, which are roughly consistent with weight values given for Boeing commercial aircraft.[63]

Table 3.15. Estimated Weight of Paint and Primer for MAF Aircraft

	Relative Wetted Area	Paint and Primer Weight (lbs)
C-130	0.345	248
C-17	1.000	720
C-5	1.392	1,002
KC-135	0.460	331
KC-10	0.862	620

Results and Conclusions

No ways to move to a primer-free process have been identified, but we hypothesize that such a process could reduce the weight of paint and primer by 50 percent. Even if these weight savings could be realized, the fuel savings would be small, only about 0.7 MG of fuel per year, as shown in Table 3.16. This translates to $2.4 million per year in savings. However, without new technology, lack of a primer would have severe consequences in the form of corrosion. Therefore, we recommend that the Air Force not adopt changes to the painting process but continue to monitor industry for improvements in paint technology.

---

[61] Boeing, "C-17 Globemaster III Technical Specifications," undated (c); United States Coast Guard, Office of Aviation Forces (CG-711), "HC-130H: Hercules," online, last modified on June 28, 2013; Günter G. Endres, *The Illustrated Directory of Modern Commercial Aircraft,* Osceola, Wisc.: MBI Publishing Co., 2001.

[62] Sherwin-Williams Aerospace Coatings, "Military Aerospace Coatings," product data, online, undated.

[63] Endres, 2001.

Table 3.16. Cost-Effectiveness of Paint Reduction

	Annualized Fuel Savings (MG / Percent)	Annualized Reduced Paint Cost ($FY13M)	Annualized Fuel Savings ($FY13M)	Break-Even Fuel Cost ($FY13M)
		Negative Additional Impacts		
C-130	0.1 (0.1%)	Minimal	0.3	Minimal
C-17	0.3 (0.1%)	Minimal	1.3	Minimal
C-5	0.0 (0.1%)	Minimal	0.2	Minimal
KC-135	0.1 (0.1%)	Minimal	0.4	Minimal
KC-10	0.1 (0.1%)	Minimal	0.2	Minimal
Total	0.7 (0.1%)	Minimal	2.4	

Microvanes

Microvanes are a passive flow-control technology that alters the flow around the aft cargo ramp of the C-130. The drag around the upswept aft ramp can account for over 10 percent of the total drag of a C-130.[64] The particularly high aft drag on C-130 has been a known issue, and research as early as the 1970s looked to deal with this problem using strakes.[65]

Description

Lockheed Martin Corporation, the manufacturer of the C-130, developed microvanes as a way to reduce the drag around the upswept after body of the C-130.[66] Although all aircraft, particularly military transport aircraft, have significant upsweep on the afterbody, the C-130 has particularly high aft drag as a result of its overall size and aft ramp design. Since the initial design of the C-130, advancements in computational fluid dynamics have allowed for refinements in microvane design and placement.[67]

Separately, Vortex Control Technologies has developed finlets for the C-130, which also reduce drag on the aft portion of the C-130.[68] These finlets are much larger than the

---

[64] Brian Smith, Patrick Yagle, and John Hooker, "Reduction of Aft Fuselage Drag on the C-130 Using Microvanes," 51st AIAA Aerospace Sciences Meeting Including the New Horizons Forum and Aerospace Exposition, Grapevine, Texas, January 7–10, 2013.

[65] Smith, Yagle, and Hooker, 2013.

[66] Smith, Yagle, and Hooker, 2013.

[67] Kyle Smith, "Fuel Efficiency Initiatives," briefing, Lockheed Martin Advanced Development Programs, undated (b).

[68] Vortex Control Technologies, "Finlets Technology—Applicable Aircraft: C130 /L100 Hercules," online, undated.

microvanes developed by Lockheed Martin. However, only six finlets are required per side,[69] or as few as four per side,[70] whereas 18 microvanes would be required.[71]

Flight test results from August 2011 suggest that microvanes can reduce fuel consumption by 2 to 3 percent,[72] whereas finlets are reported to reduce fuel use by 7 percent.[73] However, computational fluid dynamic results on finlets show a drag reduction similar to the drag reduction of microvanes.[74] Previous work by Lockheed Martin indicates that large strakes, while offering additional drag reduction, are likely to interfere with airdrop operations.[75]

Analytic Approach

For the purposes of this analysis, we assume the average of the C-130 microvane flight test results, which gives a total fuel reduction of 2.5 percent. This savings is also consistent with published computation fluid dynamic results of finlets. Costs for microvanes have not been published; however, we expect them to be cheaper than finlets, which are listed at $394,000 uninstalled.[76] If the Air Force procured them for the entire fleet of 321 C-130s— the fleet size given in FY 2013 PB for FY 2014—the cost per aircraft would likely decrease. Using these assumptions, we estimate the cost for microvanes to be $250,000 per aircraft. However, given the uncertainty about this cost, we perform a sensitivity analysis to derive the maximum cost at which microvanes would be cost-effective.

Results

We find that a 2.5 percent decrease in fuel usage for the C-130 fleet yields a modest annualized savings of around 0.6 MG of fuel, which corresponds to $2.2 million of annualized fuel savings. Recall that the 0.6 MG is an annualized fuel savings that takes into account the retirement of the C-130 fleet, meaning that the fuel savings from microvanes will not be realized indefinitely. At $250,000 per aircraft, the total cost to retrofit the entire fleet of 321 aircraft would be slightly more than $80 million. Using the real discount rate of 1.1 percent, this yields an annualized cost of about $873,000, significantly less than the

[69] Consulting Aviation Services, "VC Finlet: Reduces Drag on Upward Swept Fuselage Aircraft," online, undated.

[70] Erdem Ayan, Hakan Telli, and Y. Volkan Pehlivanoglu, *Computational Investigation of C-130 Afterbody Drag Reduction by Finlets*, Istanbul, Turkey: Aeronautics and Space Technologies Institute, Turkish Air Force Academy, undated.

[71] Smith, undated (b).

[72] Graham Warwick, "Lockheed Developing Winglets for C-130, C-5," *Aerospace Daily & Defense Report*, October 6, 2011, p. 3.

[73] Vortex Control Technologies, *2013 Program Price List*, January 2013.

[74] Ayan, Telli, and Pehlivanoglu, undated; Smith, undated (b).

[75] Jeff Rhodes, "Tweak My Ride," *Code One* (Lockheed Martin), March 7, 2012.

[76] Vortex Control Technologies, undated.

$2.2 million in annualized fuel savings. In fact, microvanes would be cost-effective at today's fuel price as long as the cost to retrofit the entire fleet was less than $198 million or $618,000 per aircraft. Judging by the finlet list price, this seems very likely.

Conclusions

Microvanes appear to be cost-effective for the C-130 fleet, as seen in Table 3.17. We recommend that the Air Force plan to install microvanes, or finlets, provided the total cost per aircraft is less than $618,000. However, before installation, the Air Force should verify no degradation to mission performance or aircraft maintainability.

Table 3.17. Cost-Effectiveness of Microvanes

	Neutral Additional Impacts			
	Annualized Fuel Savings (MG / Percent)	Annualized Microvanes Cost ($FY13M)	Annualized Fuel Savings ($FY13M)	Break-Even Fuel Cost ($FY13M)
C-130	0.6 (0.6%)	0.9	2.2	$1.51
C-17	N/A	N/A	N/A	N/A
C-5	N/A	N/A	N/A	N/A
KC-135	N/A	N/A	N/A	N/A
KC-10	N/A	N/A	N/A	N/A
Total	0.6 (0.1%)	0.9	2.2	

Ground Towing

Aircraft can be towed to and from the parking ramp and the EOR in lieu of operating the engine during launch and recovery, significantly reducing fuel consumption. However, this requires a considerable investment in logistics.

Description

Aircraft typically taxi to and from the runway under the power of one or more engines. Aircraft engines are not optimized for taxiing and therefore this is an inefficient way to move aircraft on the ground. However, there is obvious simplicity to using the aircraft's own engines.

MAF aircraft typically start engines on the parking ramp during the launch sequence and then taxi to the EOR for takeoff. During landing, aircraft return from the EOR to the parking ramp under engine power, and then the engines are shut down. If aircraft were instead towed to and from the EOR, before starting engines, a significant amount of fuel could be saved, since the fuel used by tow vehicles and any additional APU use is nominal when compared to aircraft engines operation.

Analytic Approach

One significant cost associated with towing aircraft is the additional maintenance personnel required. We estimate that it takes three hours per sortie to tow aircraft to and from the EOR. For each direction, this gives 30 minutes for maintenance personnel to prepare for the tow, such as assembling a team, retrieving a tow vehicle and other equipment; 30 minutes to tow an aircraft to or from the EOR; and 30 minutes to return tow vehicle and other equipment and transition tow team members to other maintenance duties.[77]

Figure 3.7 depicts the basic implementation of towing. During an aircraft launch sequence, the aircraft would be towed to the EOR, most likely after the aircraft doors were closed and before the engines were started. Subsequently, the engines would be started at the EOR. During the recovery process the aircraft would shut down at the EOR and be towed to the parking ramp. The towing to and from the EOR would require an additional nine to 11 maintenance man-hours depending on the MDS. The analysis approximates normal aircraft taxiing to be 30 minutes per sortie and towing to be one hour per sortie (not including preparation time before the tow or time to return the equipment time after the tow) to tow the aircraft to and from the EOR, thereby adding an extra 30 minutes to each sortie.

The total number of man-hours required to tow an aircraft to and from the EOR is estimated to be nine to eleven, depending on MDS, and detailed in Table 3.18.[78] To

Figure 3.7. Ground Towing Diagram

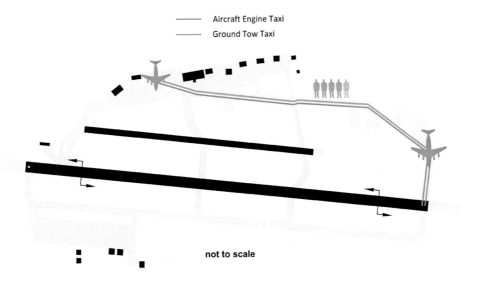

——— Aircraft Engine Taxi
——— Ground Tow Taxi

not to scale

---

[77] This estimation of 1.5 hours to complete a tow is slightly higher than those produced by the Logistics Composite Model for any of the MAF MDSs.

[78] This information was received from AMC/A4MS on July 8, 2013.

Table 3.18. Maintenance Manpower Requirements

	Maintenance Personnel to Launch and Recover Aircraft	Maintenance Personnel to Tow Aircraft [1]	Maintenance Personnel Time per Sorite	Total Additional Man-Hours per Sortie
C-130 KC-135 KC-10	2	5 – tow team supervisor, tractor operator, two wing walkers, tail walker	2 – launch/recovery team for one additional hour / 3 – additional personnel for three hours	11
C-17	3	5 – tow team supervisor, tractor operator, two wing walkers, tail walker	3 – launch/recovery team for one additional hour / 2 – additional personnel for three hours	9
C-5	3	5 – tow team supervisor, tractor operator, two wing walkers, tail walker [2]	3 – launch/recovery team for one additional hour / 2 – additional personnel for three hours	9

[1] Assumes aircrew are onboard performing as brake operator
[2] Assumes aircrew flight engineer onboard

SOURCE: Data provided to the authors by AMC/A4MS, July 8, 2013.

remain conservative, we assumed a tail walker (i.e., an individual trailing the aircraft) will be required, although tail walkers are required only when backing up or moving into or out of a hangar and may not be necessary for tows along parking ramps, taxiways, and EORs. Further, fewer man-hours may be required if towing became a standard part of launch and recovery procedures; we estimate that the requirement could be reduced by as much as three man-hours for all MDS. To calculate the cost to conduct a tow, we used a Staff Sergeant (E-5) as the average rank. In accordance with Comptroller direction, the analysis assumed an annual DoD composite hourly rate for an E-5 in the Air Force of $42.92 for FY 2013.[79] To acknowledge that ground towing may not always be feasible, especially at dual-use airfields, we estimated that towing could be done for only 80 percent of total sorties.

---

[79] John P. Roth, Office of the Under Secretary of Defense (Comptroller), "FY 2013 Department of Defense (DoD) Military Personnel Composite Standard Pay and Reimbursement Rates," memorandum, April 9, 2012.

Results and Conclusions

Table 3.19 shows that towing all MDS except the C-130 and KC-135 would be cost-effective at current fuel prices, despite the additional logistical costs. The overall fuel savings is approximately 27 MG of fuel; if the savings associated with C-130 and KC-135 were excluded, the total fuel savings would be only 18 MG. Since both the C-130 and KC-135 burn comparatively less fuel during taxiing, the costs associated with towing are not offset by the possible fuel savings. The table does not account for the costs associated with the increased fuel use required to operate the tow vehicle or APU during tow, but that cost is small relative to the overall fuel savings from the reduced engine operations. Although towing may be a cost-effective option, there may be unforeseen issues associated with congestion at the EOR, as part of either the launch or recovery procedure or an aircraft malfunction. Additional analysis is recommended to study airfield management considerations for an increase in operations and maintenance at the EOR.

To increase maintenance and airfield flexibility, an option could be to perform the entire launch and recovery procedures near the EOR, prepositioning the aircraft and equipment before the launch or recovery procedure. Although the logistics costs are about the same, this approach provides more flexibility by allowing tows to take place any time before or after the launch or recovery sequence and maintains the benefits of reduced engine operation. A mitigating factor, when possible, may be to combine the tow out of one aircraft with the tow in of another. Weather and other considerations, such as space availability, would play a factor in feasibility, and airfield management issues would require research at each location.

Another method that could reduce engine taxiing comes from commercial technology, for example, the Electric Green Taxiing System. This system uses the APU to power motors mounted to the main landing gear wheels thereby allowing aircraft to taxi without requiring the use of aircraft engines. Each main wheel is equipped with an electric motor,

Table 3.19. Cost-Effectiveness of Ground Towing

	Negative Additional Impacts			
	Annualized Fuel Savings (MG / Percent)	Annualized Ground Towing Cost ($FY13M)	Annualized Fuel Savings ($FY13M)	Break-Even Fuel Cost ($FY13M)
C-130	4.6 (5.0%)	48.9	17.2	$10.63
C-17	13.9 (2.7%)	30.2	51.7	$2.18
C-5	2.2 (2.3%)	3.6	8.1	$1.66
KC-135	4.1 (2.2%)	19.3	15.2	$4.73
KC-10	2.0 (1.8%)	4.7	7.4	$2.36
Total	26.7 (2.7%)	106.6	99.6	

reduction gearbox, and clutch assembly to drive the aircraft. Power electronics and system controllers give pilots total control of the aircraft's speed, direction, and braking during taxi operations.[80]

Lift Distribution Control

Lift distribution control is the active management of the lift position along the wing. Such a system was designed and implemented on the C-5 fleet to reduce wing fatigue.[81] A similar system has been proposed for the C-130, and studies show that in addition to removing the need for wing-relieving fuel, it also changes the angle of attack of the aircraft, which reduces drag.[82]

Description

Basic lifting line theory tells us that the optimum spanwise distribution of lift has an elliptical profile.[83] However, other constraints and considerations mean that actual aircraft wings have more complex profiles. Specifically, given weight constraints, other more preferred lift distributions exist.[84] Using active techniques, such as a proposed C-130 lift distribution control system, the lift distribution can be varied throughout flight.[85]

The C-130 currently requires wing relief fuel above a certain weight to maintain the structural integrity of the wing under high loads.[86] This relief fuel cannot be burned inflight, it increases the weight of the aircraft, and it decreases available fuel quantities. Lockheed's proposal for a lift distribution control system (LDCS) sought to eliminate the need for this fuel by allowing the ailerons to deflect symmetrically, therefore bringing the center of lift inboard and reducing wing bending loads.[87] Lockheed notes in its analysis that the use of LDCS also serves to change the angle of attack of the aircraft. This change in

---

[80] This is a product being developed in partnership between Honeywell and Safran. See Honeywell and Safran Aerospace Defence Security, "Electric Green Taxiing System," brochure, December 2011.

[81] T. E. Disney, "C-5A Active Load Alleviation System," *Journal of Spacecraft and Rockets*, Vol. 14, No. 2, 1977, pp. 81–86.

[82] Kyle Smith, "C-130 Hercules Fuel Efficiency, Initiatives," Lockheed Martin Advanced Development Programs, undated (a); Smith, undated (b).

[83] R. T. Jones and T. A. Lasinski, "Effect of Winglets on the Induced Drag of Ideal Wing Shapes," Washington, D.C.: National Aeronautics and Space Administration, NASA Technical Memorandum 81230, September 1980.

[84] L. Prandtl, "Über Tragflügel des Kleinsten Induzierten Widerstandes," *Zeitschrift für Flugtechnik und Motorluftschiffahrt*, Vol. 24, 1933, pp. 305–306.

[85] Smith, undated (a).

[86] Cynthia Dion-Schwarz et al., *FCS Vehicle Transportability, Survivability, and Reliability Analysis*, Alexandria, Va.: Institute for Defense Analysis, April 2005.

[87] Smith, undated (a).

angle of attack can reduce total drag on the aircraft. The mechanism of drag reduction is similar to that of microvanes.

Because of the similarity in the drag reduction mechanism of microvanes and LDCS, these improvements are likely not additive. Further testing would need to be done to quantify the benefits of implement microvanes and LDCS.

Analytic Approach

Lockheed's analysis of LDCS suggest that the drag count reduction is about half that of microvanes in most of the flight regime.[88] From the microvane analysis, we therefore estimate the net fuel reduction resulting from LDCS to be 1.25 percent. Although not a direct parallel, the Air Force by 1977 had installed an active lift distribution on 77 C-5As for $15.5 million in then-year dollars. [89] This translates to approximately $50 million in FY 2013 dollars for the entire fleet, or about $650,000 per aircraft. Scaling this simply by operating empty weight of the aircraft gives an LDCS cost estimate of $200,000 per aircraft.

We did not quantify the benefit of being able to carry more cargo or fly farther as a result of the elimination of the wing relief fuel. First, this does not directly increase fuel efficiency, and, second, our analysis of GDSS data indicates that wing relief fuel is needed on only a very small portion of flights.

Results and Conclusions

LDCS appears to be cost-effective but yields total fuel savings only half that of microvanes at similar cost. Table 3.20 shows that the total annualized fuel savings from LDCS is $1.1 million per year. We therefore recommend that the Air Force consider microvanes over LDCS. However, an analysis should be conducted to quantify the benefits of installing both microvanes and LDCS. The operational benefits of LDCS appear minimal and are likely not significant enough to drive the decision to install LDCS.

---

[88] Smith, undated (a).

[89] United States General Accounting Office, *C-5A Wing Modification: A Case Study Illustrating Problems in the Defense Weapons Acquisition Process*, Washington, D.C., March 1982.

Table 3.20. Cost-Effectiveness of Lift Distribution Control

	Positive Additional Impacts			
	Annualized Fuel Savings (MG / Percent)	Annualized LDCS Cost ($FY13M)	Annualized Fuel Savings ($FY13M)	Break-Even Fuel Cost ($FY13M)
C-130	0.3 (0.3%)	0.7	1.1	$2.42
C-17	N/A	N/A	N/A	N/A
C-5	N/A	N/A	N/A	N/A
KC-135	N/A	N/A	N/A	N/A
KC-10	N/A	N/A	N/A	N/A
Total	0.3 (0.0%)	0.7	1.1	

4. Cost-Ineffective Options for Reducing Fuel Use

The previous chapter discussed the options for reducing fuel that yielded net cost savings. This chapter describes options that reduce fuel consumptions but are not cost-effective. The sections follow the same format as the previous chapter: description, analytic approach, results, and conclusions.

Engine Modification or Replacement

A potentially less-costly fuel saving alternative to fleet recapitalization is to improve the engine performance of existing aircraft, either by re-engining or modifying existing engines. In cases where a modern engine is already commercially available and can be substituted for the older current engine with minimal modification, such a replacement is worth evaluating. Similarly, engine technology is continually advancing, and where new components can replace the old without a prohibitive cost, cost-effective fuel saving opportunities may be found.

Description

Generally, designing a new engine from scratch will not be cost-effective for the Air Force. The savings accumulated by the Air Force's small fleets and low flying hours cannot offset the substantial development cost, which the Air Force would bear entirely in the case of a custom-built engine. Therefore, a practical re-engining program would require identifying an existing engine in the same thrust class that can replace an existing engine with minimal modifications to either the engine or the airframe. Not surprisingly, the options tend to be limited. In the case where such an engine exists, the efficiency gains can be substantial, as will be seen in the analysis results.

An alternative to re-engining is to modify or refurbish the existing engines, incorporating new technology in the form of new component designs, high performance materials, and other improvements. The efficiency gains will generally be smaller than what can be achieved with a new modern engine, but the program cost is also smaller.

Analysis Approach

We base our work on the cost-effectiveness analysis the National Research Council (NRC) published in 2007 looking at a variety of engine options.[1] In particular, the council considered new engines and engine upgrades for each of the MAF aircraft and evaluated the associated costs

---

[1] Committee on Analysis of Air Force Engine Efficiency Improvement Options for Large Nonfighter Aircraft, National Research Council, *Improving the Efficiency of Engines for Large Nonfighter Aircraft,* Washington, D.C.: The National Academies Press, 2007.

and fuel savings. As noted, NRC's results do not constitute a complete cost assessment, as it did not monetize all aspects of such a program, such as residual value of existing engines.

For the purposes of our analysis, we updated the results of the NRC study to be consistent with our assumptions. This includes higher fuel cost and a lower discount rate, which will tend to make fuel savings programs more cost-effective. We also adjusted cost and fuel savings assumptions in the NRC work where appropriate, as we will discuss. Similarly, we applied the retirement schedule of the aircraft to derive NPV savings. We consider only the cost of the engines and the savings from reduced fuel use. We do not attempt to evaluate improvements in maintenance costs, aircraft life, or any other costs or benefits associated with re-engining or engine modification. In this way, the results shown here are a preliminary estimate of whether an approach is worthy of more detailed consideration.

C-130H re-engining. The Allison T56 engines on the C-130H could be replaced with the Rolls-Royce AE 2100 engine used on the C-130J. The NRC considered several potential new engines: We focus on the AE 2100 because it gives the greatest fuel savings. We adjusted the numbers in the NRC report in two ways (apart from being inflated to FY 2013 dollars and applying our assumptions). First, in the NRC report, the cost of the engines themselves represented 93 percent of the total program cost, which we found to be inconsistent with prior re-engining programs using existing engines, where the engine cost is roughly 56 percent of the total program cost.[2] We use the latter percentage in our calculations and therefore arrive at a significantly higher, and we believe more accurate, total program cost. Second, the annual fuel savings is estimated at 28 percent in the NRC report. Our assessment suggested that this was too optimistic an assessment, and we use a fuel savings of only 8 percent per flight hour.[3] Needless to say, these two changes have a dramatic effect on the conclusions regarding the potential cost-effectiveness of a re-engining program.

KC-10 engine modifications. The NRC report evaluated engine modifications for the KC-10, which involved incorporating a wide range of new technologies into the engine system. For these modification programs, we use the costs and fuel-efficiency improvements reported by the NRC, updated to FY 2013 dollars and our baseline assumptions.

---

[2] Fifty-six percent derived from data in the FY 2006 President's Budget P-3A Exhibit for the "C-135 Reengine MN-3009E" modification. Total cost includes structural and systems modifications needed to support new engines and performance requirements. Department of the Air Force, *United States Air Force Committee Staff Procurement Backup Book FY 2006/2007 Budget Estimates: Aircraft Procurement Air Force,* Vol. II, Washington, D.C.: SAF/FMB, February 2005, pp. 580–583.

[3] Lockheed has stated that "Four Rolls-Royce AE 2100D3 engines, each flat rated at 4,591 shaft horsepower, generate 29 percent more thrust although they are 15 percent more fuel efficient" (see Lockheed Martin, "C-130J Super Hercules Worldwide Fleet Soars Past 1 Million Flight Hours," Marietta, Ga., May 14, 2013). A 15 percent fuel savings per mile translates into an annual savings of 8 percent over the total hours flown by the C-130H fleet. Recall here that we keep total flight hours constant.

The KC-135 fleet will be upgraded with a CFM International engine as part of a propulsion upgrade program,[4] and for other MAF aircraft, no viable near-term re-engining options have been identified. The C-5 Reliability Enhancement and Re-Engining Program is ongoing, and the C-130J already has a modern engine. Engines in the same thrust class and of similar size as the KC-10 and C-17 do not exist, although the Air Force Research Laboratory is studying potential future engine options for the C-17.

Results

The program cost and fuel saving estimates for each of the above programs are shown in Table 4.1. We find that none of these programs are cost-effective. Although we do not consider the benefits of these programs beyond fuel savings, the costs are so much greater than any increase in fuel efficiency that it is unreasonable to expect these other benefits to tip the balance and make the programs cost-effective overall. The modest annual fuel savings simply do not come close to recouping the program cost over the remaining life of the aircraft.

Table 4.1. Costs and Fuel Savings Associated with Re-Engining and Engine Modification Programs Evaluated

	Total Cost ($FY13M)	Fuel Savings (percent)
C-130H re-engining	2,300	8
KC-10 engine modification	40.5	0.3

Conclusions

The annualized results are shown in Tables 4.2 and 4.3. The breakeven fuel cost for these programs is in most cases well over $20 per gallon. Even for the KC-10 engine modification, which looks inexpensive by comparison, the fuel price would have to increase more than 40 percent for the program to break even. As noted above, many other potential benefits to these programs are not captured here, which is why we assess that these programs have positive additional impacts.

---

[4] Maj. Keither Bland, "KC-135 Upgrades to Keep Aircraft Flying for Decades," Air Mobility Command Public Affairs, August 9, 2012.

Table 4.2. Cost-Effectiveness of Re-Engining

	Positive Additional Impacts			
	Annualized Fuel Savings (MG / Percent)	Annualized Re-engining Cost ($FY13M)	Annualized Fuel Savings ($FY13M)	Break-Even Fuel Cost ($FY13M)
C-130	1.0 (1.0%)	24.9	3.6	$25.97
C-17	N/A	N/A	N/A	N/A
C-5	N/A	N/A	N/A	N/A
KC-135	N/A	N/A	N/A	N/A
KC-10	N/A	N/A	N/A	N/A
Total	1.0 (0.1%)	24.9	3.6	

Table 4.3. Cost-Effectiveness of Engine Modifications

	Positive Additional Impacts			
	Annualized Fuel Savings (MG / Percent)	Annualized Engine Mods Cost ($FY13M)	Annualized Fuel Savings ($FY13M)	Break-Even Fuel Cost ($FY13M)
C-130	N/A	N/A	N/A	N/A
C-17	N/A	N/A	N/A	N/A
C-5	N/A	N/A	N/A	N/A
KC-135	N/A	N/A	N/A	N/A
KC-10	0.1 (0.1%)	0.4	0.3	$5.24
Total	0.1 (0.0%)	0.4	0.3	

Winglets

Winglets are wingtip devices designed to improve the L/D of an aircraft. As the span of a wing increases, the induced drag decreases, approaching zero in the limit of an infinite span.[5]

Of course, there are many practical limitations with very large span wings, including increased parasitic drag, increased weight of the wing, and parking and operational constraints.[6] Winglets increase the apparent span of the wings without many of the associated negative consequences.[7]

---

[5] National Advisory Committee for Aeronautics, *The Spanwise Distribution of Lift for Minimum Induced Drag of Wings Having a Given Lift and a Given Bending Moment,* Moffett Field, Calif.: Ames Aeronautical Laboratory, Technical Note 2249, December 1950.

[6] Raymer, 2006.

[7] George Larson, "How Things Work: Winglets," *Air & Space Magazine*, September 2001.

Description

Significant efforts to design and measure the effectiveness of winglets began with Whitcomb in the 1970s.[8] Whitcomb succinctly describes winglets as a wing-tip extension that yields greater reduction in drag than a simple wing extension with the same structural weight.[9]

Flow visualization of Whitcomb's winglet design in Figure 4.1 shows that it creates a much smaller vortex than a simple fairing.[10] The smaller vortices generated at the wingtips have less energy, because a portion of the energy from these vortices is captured by the winglets to generate additional forward thrust.[11] In simple terms, the vortex is weaker because the winglet prevents the spillage of air at the wingtip from the high pressure region below the wing to the low pressure region above the wing.

Because of the drag reduction offered by winglets, commercial airlines began retrofitting large numbers of aircraft beginning around 2003 when fuel prices were approximately $1 per

Figure 4.1. Smoke Visualization of Flow over a Simple Fairing and over Whitcomb's Winglet

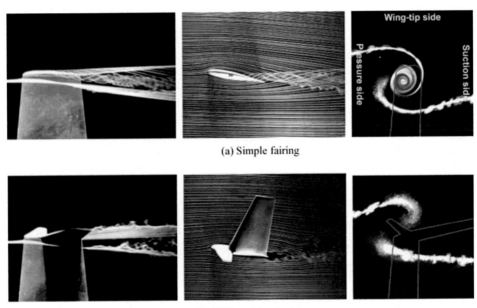

(a) Simple fairing

(b) Whitcomb's winglet

---

[8] Richard T. Whitcomb, *A Design Approach and Selected Wind-Tunnel Results at High Subsonic Speeds for Wing-Tip Mounted Winglets,* Hampton, Va.: Langley Research Center, NASA Technical Note D-8280, July 1976.

[9] Clayton J. Bargsten and Malcolm T. Gibson, *NASA Innovation in Aeronautics: Select Technologies That Have Shaped Modern Aviation*, Washington, D.C.: National Aeronautics and Space Administration, NASA/TM-2011-216987, August 2011.

[10] Myong Hwan Sohn and Jo Won Chang, "Visualization and PIV Study of Wing-Tip Vortices for Three Different Tip Configurations," *Aerospace Science and Technology*, Vol. 16, No. 1, January–February 2012, pp. 40–46.

[11] NASA, "Winglets Save Billions of Dollars in Fuel Costs," online, undated.

gallon in FY 2013 dollars.[12] As fuel prices have risen, retrofits have continued. Currently, United Airlines is replacing existing winglets with more advanced winglets to gain the marginal improvement they offer.[13] Although commercial airlines have been retrofitting their aircraft with winglets, none of the MAF aircraft have been retrofitted, and only the C-17 was delivered with winglets installed.[14] It appears that the C-17 will remain the only aircraft in the fleet with winglets, as current plans for the KC-46A do not include winglets.[15]

Analytic Approach

In 2012, RAND published work that characterized the cost-effectiveness of adding winglets to the Air Force tanker fleet, specifically the KC-135 and KC-10.[16] The conclusion of this work was that the addition of winglets to the tanker fleet is not cost-effective. This RAND work computed the cost-effectiveness of winglets for the tanker fleet through detailed flight modeling and cost estimations. However, the conclusion can be understood very simply; if we assume that, on average, commercial carriers have correctly chosen when to install winglets, we can use their decision point as an analogy for the Air Force, putting other differences aside. Figure 4.2 shows the approximate price of fuel from 2000 to 2008 in FY 2013 dollars along with the approximate time airlines began retrofitting their fleets. Many of the retrofits occurred around the $1.50 per gallon price. Depending on the estimate, commercial airlines may fly as many as 4,400 hours per aircraft per year.[17] According to the FY 2013 PB programmed flight hours for FY 2014, aircraft in the MAF fleet are expected to fly between slightly less than 300 to slightly more than 800 flight hours per year, depending on MDS.[18] A simple calculation then tells us that we might expect winglets to be cost-effective for the MAF fleet starting at $8.25 per gallon all the way up to $22 per gallon, depending on the MDS.

For our detailed analysis, we use the winglet modification costs for the KC-10 and KC-135, as presented in the RAND work on winglets for the USAF tanker fleet, with dollar amounts adjusted for inflation. We then performed a regression for winglet kit cost including installation as well as research, development, test, and evaluation (RDT&E) as a function of aircraft operational empty weight (OEW). Winglet cost data came from Aviation Partners Boeing

---

[12] Norton, 2012.

[13] United Airlines, "United Airlines is First to Install Split Scimitar Winglets," press release, July 17, 2013.

[14] Scott Schonaeuer, "'Winglets' Could Save Air Force Millions on Fuel," *Stars and Stripes*, October 1, 2007.

[15] Amy Butler, "KC-46A—First Photos . . . Sans Winglets," Ares, *aviationweek.com* blog, April 7, 2011.

[16] Norton, 2012.

[17] Airbus, "The A330/A340 Family Jetliners Benefit from Lower Maintenance Costs," press release, April 16, 2009.

[18] United States Department of Defense, *Fiscal Year (FY) 2013 President's Budget: Flying Hour Program (PA)*, an extract of the Programmed Data System (PDS), Washington, D.C., February 2012, not available to the general public.

Figure 4.2. Airline Winglet Retrofits over Time Based on Fuel Price

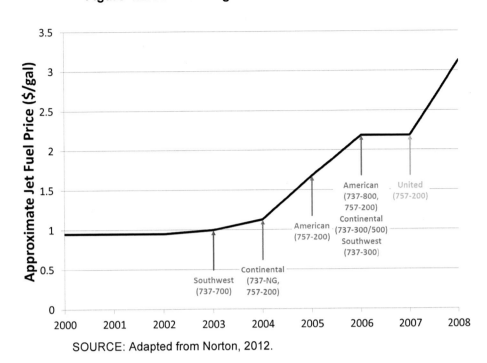

SOURCE: Adapted from Norton, 2012.

and from *Aviation Week*.[19] We derived the following regression for winglet kit cost, $P_{kit} = 3.475W_{empty}^{1.108}$, where P_{kit} is the uninstalled cost of the kit in FY2013 dollars and W_{empty} is the operating empty weight of the aircraft in pounds. The OEW came from the aircraft manufacturers.[20] In addition, we assume that the kit makes up 51.5 percent of the total unit procurement cost, including installation, based on cost ratios in the previous RAND analysis. Performing a similar regression on RDT&E, we find that $P_{RDT\&E} = 0.853W_{empty}^{0.392}$, where $P_{RDT\&E}$ is the RDT&E costs in millions of FY 2013 dollars. The cost regressions are subject to error and an actual development program is likely to have costs that are significantly different; however, this is important only if winglets prove to be marginally cost-effective.

Using the previous RAND research, we assume a fuel use reduction of 4 percent for the KC-135 and 3 percent for the KC-10. Although a precise estimate would require detailed aerodynamic analysis, we simply assume that fuel use reductions for the C-130 and C-5 will be similar and estimated those at 3.5 percent. We do not include the C-17 in this analysis, since this aircraft already has winglets, and the marginal gains of installing redesigned winglets will likely prove less cost-effective than adding winglets to the other MAF aircraft. These costs and associated fuel savings are summarized in Table 4.4.

---

[19] Aviation Partners Boeing, "Blended Winglets: Improved Takeoff Performance," undated (a); David Esler "Question for Efficiency Driving Winglet Retrofits," *Aviation Week*, December 1, 2008.

[20] Boeing, "Airplane Characteristics for Airport Planning," online, undated (b); Cessna Aircraft Company, *Citation X: Specification & Description,* Units 750-0501 to TBD, , Revision B, Preliminary Wichita, Kan., March 2013; P. Jackson, ed., *Jane's All the World's Aircraft,* Jane's Information Group, 2012.

**Table 4.4. Costs and Associated Fuel Savings
of Winglets**

	Unit Procurement ($FY13M)	RDT&E ($FY13M)	Fuel Reduction (Percent)
C-130	1.7	70	3.5
C-5	10.3	131	3.5
KC-135	4.5	77	4.0
KC-10	10.0	109	3.0

Results and Conclusions

Although some analyses have concluded that winglets may be cost-effective,[21] our findings suggest that they are not. Changes in assumptions may change the relative order of effectiveness, but it is clear from Table 4.5 that winglets for the MAF fleet are not cost-effective unless fuel prices nearly double. These results should not be surprising; a simple analysis using commercial airline behavior suggested that winglets would not be cost-effective until about $8.25 per gallon. Again, note that the annualized fuel savings for each of the aircraft shown in the table is less than the yearly fuel savings associated with winglets. As discussed above, this is because these aircraft will retire, and the benefit from winglets retrofits is not indefinite. Although some additional benefits beyond fuel savings might occur, such as improved takeoff performance,[22] these benefits are modest, and we have therefore summarized the net additional effects of winglets as neutral.

Table 4.5. Cost-Effectiveness of Winglets

	Neutral Additional Impacts			
	Annualized Fuel Savings (MG / Percent)	Annualized Winglets Cost ($FY13M)	Annualized Fuel Savings ($FY13M)	Break-Even Fuel Cost ($FY13M)
C-130	0.7 (0.8%)	5.2	2.7	$7.03
C-17	N/A	N/A	N/A	N/A
C-5	0.9 (0.9%)	7.3	3.3	$8.09
KC-135	0.5 (0.2%)	6.0	1.7	$13.07
KC-10	0.9 (0.8%)	7.6	3.4	$8.35
Total	3.0 (0.3%)	26.0	11.2	

---

[21] National Research Council, Committee on Assessment of Aircraft Winglets for Large Aircraft Fuel Efficiency, *Assessment of Wingtip Modifications to Increase the Fuel Efficiency of Air Force Aircraft,* Washington, D.C.: The National Academies Press, 2007.

[22] Aviation Partners Boeing, undated (a); National Research Council, 2007.

Riblets

Riblets are grooves or protrusions on the aircraft skin designed to reduce viscous drag. Riblets achieve drag reduction by reducing the exchange of momentum within a turbulent boundary layer. This reduced drag of course leads to improved fuel efficiency.

Description

Figure 4.3 shows an example of riblets manufactured by 3M Company.[23] These riblets are isosceles triangles with a spacing of 2.8 mil (1 mil = 0.001 in.) and a height of 2.8 mil. The relatively small size of the riblets is essential for drag reduction because larger riblets will actually increase drag.[24] Although the riblets shown here, and most riblets tested experimentally, have a triangular cross-section, this is not a requirement. In fact,

Figure 4.3. Riblets Manufactured by 3M Corporation Viewed Under an Electron Microscope

SOURCE: Mouton and Graham, 2000.

---

[23] F. J. Marentic and T. L. Morris, "1986 Drag Reduction Article," Minnesota Mining and Manufacturing Co. (assignee), U.S. Patent Number 5,133,516, St. Paul, Minn., July 28, 1992.

[24] Ricardo Garcia-Mayoral and Javier Jimenez, "Drag Reduction by Riblets," *Philosophical Transactions of the Royal Society A*, Vol. 369, No. 1940, March 7, 2011, pp. 1412–1427.

configurations other than triangles may be optimal.[25] Further, research suggests that the riblets do not even need to be straight; sinusoidal riblets may be even more beneficial.[26]

NASA published early work on riblets in 1983.[27] By 1987, the effectiveness of riblets had been well established.[28] A 1987 paper presented flight test results of 3M riblets on a T-33 and showed a reduction in skin friction drag of 6 to 7 percent.[29] Flight tests in 1986, again using the 3M riblets but on a Lear Jet Model 28/29, showed similar reductions in skin friction drag. Flight testing continued in 1989 on a much larger scale, when an Airbus A320 was outfitted with 3M riblets covering 70 percent of the aircraft surface.[30] These flight tests indicated slightly less than a 2 percent reduction in total drag. In 1996, Cathay Pacific tested riblets on an Airbus A340-300 and also found an overall drag reduction around 2 percent.[31] These tests concluded that given the fuel prices at the time and accounting for installation and maintenance of the riblets the benefits were marginal.

Currently, Lufthansa Technik is developing a lacquer system to apply riblets directly as part of a multifunctional aircraft coating.[32] The hope is that such a process could increase the durability of the riblets. However, unlike the previous A340 riblet test where large sections of the aircraft were covered with riblet film, only small 10 cm × 10 cm areas are currently being studied.

Analytic Approach

Using information from the previous flight tests, which showed a 2 percent reduction in total drag, and given the uncertainties surrounding future riblet designs, we assume that future riblet designs will also achieve a 2 percent reduction in total fuel use. Further, we use the Cathay Pacific A340 flight tests to derive an estimate of riblet costs. Those tests showed little if any net

---

[25] D. W. Bechert, M. Bruse, W. Hage, J. D. T. Van der Hoeven, and G. Hoppe, "Experiments on Drag-Reducing Surfaces and Their Optimization with an Adjustable Geometry," *Journal of Fluid Mechanics*, Vol. 338, 1997, pp. 59–87.

[26] Christopher Mouton and Stephen L. Graham, "Investigation of Drag Reduction in Turbulent Flow over Riblet Covered Surfaces," paper prepared for AIAA Student Conference, Region IV, Houston, TX, 2000; and Yulia Peet and Pierre Sagaut, "Turbulent Drag Reduction Using Sinusoidal Riblets with Triangular Cross-Section," American Institute of Aeronautics and Astronautics Paper AIAA-2008-3745, prepared for 38th AIAA Fluid Dynamics Conference and Exhibit, Seattle, Washington, June 23–26, 2008.

[27] Michael J. Walsh, "Riblets as a Viscous Drag Reduction Technique," *AIAA Journal*, Vol. 21, No. 4, 1983, pp. 485–486.

[28] S. P. Wilkinson et al., "Turbulent Drag Reduction Research at NASA Langley: Progress and Plans," *International Journal of Heat and Fluid Flow*, Vol. 9, No. 3, September 1988.

[29] J. D. McLean, D. N. George-Falvy, and P. P. Sullivan, "Flight-Test of Turbulent Skin Friction Reduction by Riblets," *Proceedings of International Conference on Turbulent Drag Reduction by Passive Means*, London: Royal Aeronautical Society, 1987, pp. 1-17.

[30] Szodruch, 1991.

[31] Warsop, 2000.

[32] Lufthansa Technik, 2012.

savings; however, fuel prices have risen approximately 160 percent since those flight tests.[33] Therefore, we assume that if C-17 aircraft flew 4,400 hours per year, consistent with commercial airline norms,[34] and if fuel prices were 38 percent of their current price, the installation of riblets would be cost-effective. Table 4.6 shows this calculation. If the C-17 fleet were to increase its hours flown per aircraft to 4,400, then a 2 percent reduction in fuel use would net 54.3 MG of fuel savings. At a reduced fuel price of $1.43 per gallon, which is 38 percent of today's $3.73, riblets would break even if the total cost to maintain riblets on the entire C-17 fleet was just under $78 million, as seen in Table 4.6. This set of assumptions yields an annual cost estimate to install and maintain riblets on a C-17 of $350,000 per aircraft. This analysis assumes that there is no marginal weight for riblets, which is to say that the weight of the riblets is approximately equal to the weight of the paint it displaces.[35]

To extend the C-17 cost results to other airframes, we estimated the relative wetted area of each aircraft, and scaled the riblet cost proportionally.[36] These estimates are shown in Table 4.7.

Table 4.6. Break-Even Riblet Cost for the C-17 Fleet Under Certain Conditions

	Fuel Reduction	Flight Hours per Year	Fuel Reduction (MG)	Fuel Savings ($FY13M) @ $1.43/gal
Notional	2.0%	4,400	54.3	77.7

Table 4.7. Estimate of Annualized Riblet Cost

	Relative Wetted Area	Riblet Cost per Aircraft ($FY13M)
C-130	0.345	0.121
C-17	1.000	0.351
C-5	1.392	0.489
KC-135	0.460	0.162
KC-10	0.862	0.303

---

[33] United States Energy Information Administration, "U.S. Gulf Coast Kerosene-Type Jet Fuel Spot Price FOB," online, release date August 28, 2013.

[34] Airbus, 2009.

[35] D. W. Bechert, M. Bruse, W. Hage, and R. Meyer, "Fluid Mechanics of Biological Surfaces and their Technological Application," *Naturwissenschaften*, Vol. 87, No. 4, 2000, pp. 157–171.

[36] Boeing, undated (c); United States Coast Guard, 2013; Endres, 2001.

Under the assumptions made here, our results suggest that riblets would now be cost-effective for airlines. However, other than the small-scale tests currently being conducted by Lufthansa, no airline is installing them. This suggests that such issues as maintainability and reliability continue to be very significant. We carry these estimates forward into the analysis recognizing that they are most likely optimistic.

Results and Conclusions

Given that the Air Force flies its mobility fleet significantly fewer hours per year than commercial airlines, and that commercial airlines are not currently installing riblets other than for testing purposes, we assume that riblets would not be cost-effective for the Air Force. As Table 4.8 shows, this is indeed the case. Despite our optimistic assumption regarding riblets and neglecting maintenance and reliability concerns, riblets are not cost-effective until fuel prices reach over $7.50 per gallon, more than twice today's price. As the table shows, even above $7.50 per gallon, but below $10 per gallon, riblets would be cost-effective only for the C-17 and the KC-10. This is essentially because these aircraft accumulate the most flying hours per year. We recommend that the Air Force not pursue riblet technology for the MAF fleet but continue to monitor industry and research labs for possible advancements.

Table 4.8. Cost-Effectiveness of Riblets

	Negative Additional Impacts			
	Annualized Fuel Savings (MG / Percent)	Annualized Riblets Cost ($FY13M)	Annualized Fuel Savings ($FY13M)	Break-Even Fuel Cost ($FY13M)
C-130	1.8 (2.0%)	38.8	6.9	$21.13
C-17	10.2 (2.0%)	77.9	37.9	$7.67
C-5	1.9 (2.0%)	33.0	7.0	$17.47
KC-135	3.8 (2.0%)	63.3	14.1	$16.72
KC-10	2.3 (2.0%)	17.9	8.4	$7.91
Total	19.9 (2.0%)	230.9	74.3	

New Aircraft

Perhaps the most obvious and expensive way to improve fuel efficiency is to purchase new aircraft. As aircraft are developed, they continue to become more fuel efficient.[37] In fact, the desire for improved fuel efficiency is to a large extent what drives the production of new

---

[37] GRID-Arendal, "Aviation and the Global Atmosphere," online, 2001.

aircraft.[38] Commercial airlines have been able to take advantage of much of these savings with fleets that average only 11.5 years old.[39]

Description

Airplane technology has continued to evolve and bring about improved fuel efficiency. An example of this is the 747-8, a follow-on to the 747-400. The -8 designation refers to the incorporation of technologies from the 787. Design improvements to the 747-8 include an improved airfoil design, raked wingtips, new GEnx engines, and use of advanced materials.[40]

Improvements in aircraft fuel efficiency over time have been significant. Measured as fuel burn per passenger-mile, fuel efficiency improved 70 percent from 1959 to 1995.[41] Some estimates anticipate further improvements in fuel efficiency of slightly more than 20 percent between 2000 and 2040.[42] These estimates show that the rate of improvement is decreasing with time. Figure 4.4 shows the normalized fuel efficiency over time using a curve fit proposed by Peeters and the data presented by Vedantham.[43]

Analytic Approach

Using the normalized fuel efficiency, we derive an estimate for fuel efficiency improvements should the current MAF fleet be recapitalized. This analysis excludes the tanker fleet, since a recapitalization is currently under way.[44] The C-130J is the newest aircraft in the fleet with the first flight occurring in 1996.[45] The C-130J is a variant of the C-130H, which was first

---

[38] Boeing, "737 MAX," *New Air Plane*, online, undated (a); Boeing, "Defining 21st Century Flight, 787 Dreamliner," online, undated (d); and Airbus, "The A320 New Engine Option," online, undated.

[39] Laurence Reid, "Will the North American Commercial Jet Fleet Experience Growth in the Next Ten Years or Will the Next Decade Be One Solely of High Replacement and Minimal Growth for This Region?" *Ascend World Wide*, April 12, 2013.

[40] Boeing, "The New 747-8," *newairplane.com*, online, undated (f).

[41] Joosung Joseph Lee, *Historical and Future Trends in Aircraft Performance, Cost, and Emissions*, dissertation, Massachusetts Institute of Technology, 2000, Cambridge, Mass: Massachusetts Institute of Technology, 2000.

[42] P. M. Peeters, J. Middel, and A. Hoolhorst, *Fuel Efficiency of Commercial Aircraft: An Overview of Historical and Future Trends*, Amsterdam, Netherlands: Peeters Advies/National Aerospace Laboratory, NLR-R-2005-669, November 2005.

[43] Peeters, Middel, and Hoolhorst, 2005; Anu Vedantham et al., "Aircraft Emissions: Current Inventories and Future Scenarios," in *Aviation and the Global Atmosphere: A Special Report of IPCC Working Groups I and III*, Cambridge, Mass.: Cambridge University Press, January 1999.

[44] Daryl Mayer, "First KC-46 Build Begins," Wright-Patterson Air Force Base, Ohio: 88th Air Base Wing Public Affairs, July 1, 2013.

[45] Lockheed Martin, "C-130J Super Hercules Worldwide Fleet Soars Past 1 Million Flight Hours," Marietta, Ga., May 14, 2013.

Figure 4.4. Normalized Fuel Efficiency over Time

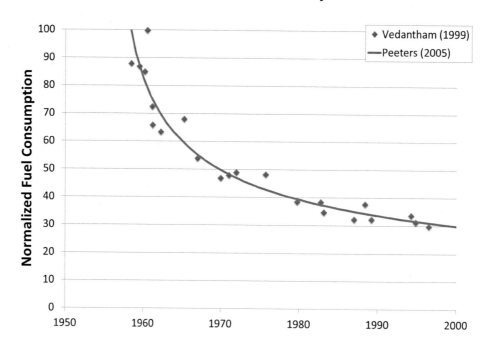

SOURCES: Adapted from Peeters, Middel, and Hoolhorst, 2005, and Vedantham et al., 1999.

introduced in 1974.[46] The C-17 had its first flight a few years before the C-130J in 1991.[47] The C-5A is a much older aircraft with the first flight in 1968.[48] Because of heritage issues, it is not possible to clearly define a baseline for any of these aircraft. For example, although the C-130J is the newest aircraft, it is still based on a legacy design and may therefore not have been able to take full advantage of new technologies. The first C-130 variant, the C-130A, was first flown in 1955.[49] Follow-on designs through the C-130J have incorporated advancements, but they have also been limited by constraints from legacy aircraft. With this in mind, we derive estimates for fuel efficiency improvement if these aircraft were recapitalized.

It is important to note that the aircraft in the current fleet will have to be recapitalized in the future because of age. This future recapitalization would yield fuel savings as well. Therefore, early recapitalization does not provide a net fuel reduction over the entire life of the aircraft.

To compute the cost-effectiveness of recapitalizing the current fleet early, we need to compute not only the fuel savings and costs associated with the early recapitalization but also the fuel savings and costs that would naturally be incurred if the aircraft were to remain in the fleet until their normal retirement date. We do this by looking at the fleet on an indefinite basis with retirements reoccurring every 45 years. The difference in fuel savings between the two options

---

[46] Lockheed Aeronautical System Company, "Meet the Hercules," *Service News*, Vol. 16, No. 1, January–March 1989.

[47] Boeing, "History: Globemaster III," online, undated (e).

[48] Lockheed Martin, "C-5 Galaxy: Heavy Lifting," online, undated (a).

[49] Federation of American Scientists, "C-130 Hercules," online, updated February 20, 2000.

can vary over time. For example, in the early retirement case, the fleet will initially be more fuel efficient than if the recapitalization did not occur. However, this fleet will also age and become less fuel efficient than then-modern aircraft. The costs associated with early recapitalization are more straightforward, since these are just the costs of shifting an RDT&E and procurement program forward in time.

We assume that in the future there will be a single strategic airlifter and a single tactical airlifter. Using an update to existing RAND cost estimating relationships,[50] we derived the RDT&E costs to be $22 billion and $4.9 billion for new-design strategic and tactical airlifters, respectively. Similarly, we determined average procurement unit costs to be $469 million and $202 million, respectively. We further assume that in real dollars these amounts remain constant and that operations and support (O&S) costs do not change.

Results

Using the regression results in Figure 4.4, we calculate the fuel reduction for the legacy fleet. These results are shown in Table 4.9. Previous work suggests that these estimates are likely optimistic.[51]

When we account for the fact that the current fleet will have to be retired in the future because of age, we see that the net savings are less than might otherwise be expected. For example, Figure 4.5 shows the relative fuel efficiency of the C-17 fleet under two assumptions. The first is based on a current estimate of the retirement schedule;[52] the second is based on a full recapitalization of the fleet in 2017. The figure shows that there are periods of time, albeit brief, where the fleet that was recapitalized earlier is actually less fuel efficient than the fleet that was recapitalized later. In this case, the net annualized fuel savings is 11 percent.

Table 4.9. Fuel Reduction of a Modern Aircraft Compared to Its Legacy Equivalent

	Fuel Reduction (Percent)
C-130H	40
C-130J	15
C-17	20
C-5	50

---

[50] Fred Timson, *Analysis of Alternatives (AoA) for KC-135 Recapitalization: Appendix E—Acquisition Costs for New-Design Alternatives*, Santa Monica, Calif.: RAND Corporation, MG-460-AF, December 2005, not available to the general public.

[51] Mouton et al., 2013.

[52] Mouton et al., 2013.

Figure 4.5. Relative C-17 Fleet Efficiency with Early Recapitalization

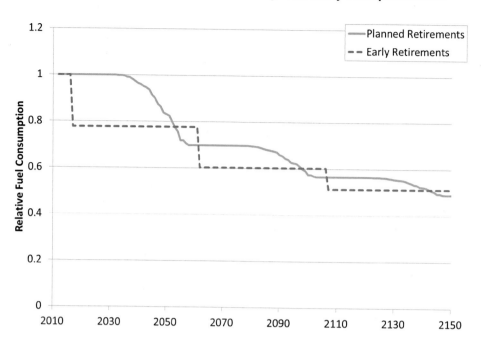

Conclusions

As expected, replacement of the current fleet with the newest technology aircraft would result in substantial fuel savings. However, the costs far exceed the potential savings. As Table 4.10 shows, most fuel savings would come from the strategic airlift fleet. However, even then fuel prices would have to reach almost $10 per gallon for it to be cost-effective for the C-17. Of course, other positive benefits are associated with new aircraft. These include better mission performance and capability, newer technologies such as improved avionics, and perhaps reduced O&S costs.

Table 4.10. Cost-Effectiveness of New Aircraft

	Positive Additional Impacts			
	Annualized Fuel Savings (MG / Percent)	Annualized New Aircraft Cost ($FY13M)	Annualized Fuel Savings ($FY13M)	Break-Even Fuel Cost ($FY13M)
C-130	13.1 (14.2%)	455.9	48.8	$34.84
C-17	55.7 (11.0%)	527.2	207.6	$9.47
C-5	19.1 (20.3%)	262.9	71.3	$13.75
KC-135	N/A	N/A	N/A	N/A
KC-10	N/A	N/A	N/A	N/A
Total	87.9 (8.8%)	1246.0	327.7	

5. Conclusions and Recommendations

Conclusions

Sixteen options for reducing fuel use in the MAF were considered. In total, 12 of these were cost-effective; however, five of these cost-effective options have significant negative implications to implementation. In addition, microvanes would be preferred to lift distribution control for reducing the drag on the aft portion of the C-130. This leaves six options that are both cost-effective and can be reasonably implemented. These options are engine-out taxiing, flying at optimum flight level and speed, basic weight reduction, APU use reduction, load-balancing improvement, and microvanes.[53]

These options are summarized in Figure 5.1. The x-axis of the figure represents how much fuel the option saves, i.e., the width of each bar gives the fuel savings of that option. The y-axis is the cost to implement the savings option. If the annualized cost of implementation is less than

Figure 5.1. Cost-Effectiveness of Fuel Reduction Options

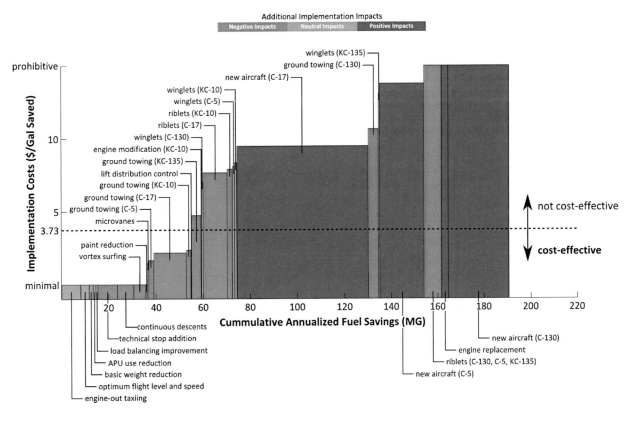

---

[53] Lift distribution control is also cost-effective and can be reasonably implemented; however, if microvanes are installed, the benefit of lift distribution control will likely be significantly reduced.

the cost of fuel, $3.73 per gallon in this case, then the option is cost-effective. As an example, ground towing of C-17 aircraft would save 13.9 MG of fuel a year and cost $2.18 per gallon.

The figure shows a couple of significant and important trends. First, of all the fuel reduction options, those that are cost-effective—below the dashed line—represent approximately 50 MG of fuel reduction a year. However, the options that are not cost-effective represent a much bigger potential for fuel reduction, about 135 MG. This illustrates that although many possibilities for fuel reduction exist, only a portion of the savings potential is cost-effective. Of the options that are cost-effective, half have significant barriers to implementation. In fact, the cost-effective options that have neutral or positive implementation impacts represent only 16 MG of fuel, or 1.6 percent of projected MAF fuel use.

Figure 5.1 is designed to show another important aspect of fuel reduction. Specifically, it answers the question of how these results change as fuel prices change by shifting the dashed line from $3.73 per gallon. However, other than negative impact options, no reasonable increases or decreases in fuel prices would change which of the options would be cost-effective. This means that the options most worth pursuing are actually independent of fuel prices, over a wide range.

Another finding we see in the figure is the clustering of options. There are four broad categories. First, a small set of options are both cost-effective and have neutral implementation impacts. Second, a larger set of options are cost-effective but have negative implementation impacts. Third, a set of options are not cost-effective and have negative implementation impacts. Fourth, a set of options are extremely not cost-effective yet have positive implementation impacts. These categories can be visualized in Table 5.1.

Table 5.1 represents a map of cost-effectiveness and implementation impacts. The top left-hand corner represents the best options, those that have positive implementation impacts and are cost-effective. The font size corresponds to the amount of savings the option provides. We see that only one option falls into the top left-hand corner, but it represents a very small savings potential. This is not unexpected, because options in the top left-hand corner would have most likely already been implemented. We see that most of the cost-effective options have neutral or negative implementation impacts. In fact, the cost-effective option with the greatest savings, ground towing, has negative implementation impacts.

The table also shows that no options are marginally cost-effective, that is to say between $3 and $5 per gallon. We also see that there are no options in the bottom right-hand corner, since such options are easily excludable from further analysis.

Table 5.1. Cost-Effectiveness of Fuel Reduction Options

	Cost-Effective (<$3/gal)	Marginally Cost-Effective ($3/gal - $5/gal)	Not Cost-Effective (>$5/gal)
Positive Implementation Impacts	Lift Distribution Control		Engine Modification **Engine Replacement** **New Aircraft**
Neutral Implementation Impacts	APU Use Reduction Basic Weight Reduction **Engine-Out Taxiing** Load Balancing Improvement Microvanes **Optimum Level & Speed**		**Riblets** Winglets
Negative Implementation Impacts	**Continuous Descents** **Ground Towing** Paint Reduction **Technical Stop Addition** **Vortex Surfing**		

Recommendations

According to the analysis presented here, we make several recommendations. We break these recommendations into two categories. The first is actions that the Air Force can take now to reduce fuel use, the second is options the Air Force should explore that have the potential to reduce fuel use in the future. Actions the Air Force should take now include the following:

Implement engine-out taxiing. Incorporating engine-out taxiing into standard operating procedures can help institutionalize engine-out taxiing and save millions of gallons of fuel annually. Engine-out taxiing places only a limited burden on operations and is done extensively in the commercial sector. Current Air Force procedures do allow for engine-out taxiing or delayed engine start when conditions permit but do not direct it.

Always fly at the optimum flight level and speed. Ensuring that crews always perform step-climbs when appropriate and adjust their speed based on their weight and altitude can save significant fuel. We understand that performing at least one step-climb when appropriate is currently standard for Air Force crews, but this should be verified. In addition, aircraft should always be flying the optimum speed for the given flight conditions. Incorporating real-time weather data, such as winds and temperature, can further increase fuel savings.

Continue to reduce basic weight of aircraft. Removing unnecessary equipment and maintaining the lowest empty weight possible can lead to modest fuel savings. The Air Force has already implemented significant efforts to reduce aircraft weight, and these efforts should be continued and reinforced.

Reduce use of the APU. Using ground equipment instead of the aircraft's APU can save over 1 MG of fuel a year. Some bases currently use the APU significantly less than others, even when temperature effects are considered. Procedures at these locations should be studied, and the lessons learned should be applied throughout the fleet. AMC should also continue to monitor and record APU use and increase the auditing of these data to identify and resolve problems early.

Ensure that loads are properly balanced. By ensuring that loads are properly balanced, fuel use could be reduced by almost 1 MG a year. The Air Force is currently improving its load-balancing procedures, and this should be reinforced and these practices should be standardized across the fleet.

Install microvanes on the C-130 fleet. Installing microvanes or finlets on the C-130 fleet could save more than half a million gallons of fuel a year. Testing is ongoing, and indications are that these can be installed easily and safely and will not degrade mission performance.

A series of additional options warrant further analysis and may prove beneficial in the future:

Expand the use of continuous descent approaches. The Air Force should examine which of its aircraft can execute continuous descent approaches and the modifications that would be required to employ this technique. Further, the Air Force should see if CDA are being executed when possible and if CDA can be expanded at military airfields. CDA could save the Air Force over 11 MG of fuel a year.

Continue testing and feasibility studies of vortex surfing. The Air Force has been testing vortex surfing with C-17s and the results are promising. This testing should continue, and detailed analysis on its implementation in the MAF fleet should be conducted. Issues such as aircraft fatigue and ride quality should be quantitatively addressed. Use of vortex surfing could save 4 MG of fuel a year for the C-17 fleet alone.

Conduct a feasibility study of ground towing. Although the negative implementation impacts associated with ground towing are clear, these may be offset by the potential to save more than 20 MG of fuel a year. Implementation of ground towing at military airfields with little traffic and long taxiing requirements may provide an excellent opportunity to test the concept. In addition, industry is developing alternatives that allow aircraft to taxi without using the engines, and these developments should be monitored.

By implementing the near-term recommendations, the Air Force could begin to save 16 MG of fuel a year, about 1.6 percent of the MAF fuel consumption, in a cost-effective manner. This could be further increased as additional options, such as CDA, vortex surfing, and ground towing, are explored. If these options could also be implemented, fuel savings could be further increased by 38 MG a year.

Fuel Reduction Options Not Included

Several options were considered but did not offer fuel savings based on the metrics used. For completeness, those topics are discussed briefly here. The discussions are meant to provide a simple rationale for not including these options, even though in most cases more robust analysis exists.

C-130H Recapitalization

The C-130J model is equipped with the more fuel efficient Rolls-Royce AE 2100D3, which offer 15 percent improved fuel efficiency over the C-130H.[1] Considering no weight change between the C-130H and C-130J and assuming that all of the increased efficiency comes from the new Rolls-Royce AE 2100D3, we can estimate the C-130J to be 15 percent more fuel efficient per mile than the C-130H.[2] However the C-130J also cruises at a higher speed than the C-130H under most conditions.[3] Since we are primarily concerned with fuel efficiency per hour, this gives an overall fuel efficiency improvement of the C-130J compared to the C-130H on a per-hour basis of slightly under 9 percent. However, by recapitalizing the C-130H with C-130J now, the first retirement of those aircraft would be pushed back from 2031 to 2062. This means that the fleet would continue to operate on a legacy platform much longer than it otherwise would. So even though the C-130J would provide initial fuel improvements to the C-130H, there is actually a much larger lost potential by not incorporating a significantly newer technology aircraft into the fleet sooner.

Engine Washing

During engine operation, the engine blades and surfaces build up deposits that make the surface finish rough. This results in inefficiencies, which can lead to increased fuel consumption, maintenance problems, and more unscheduled engine removals. Engine washing is a scheduled preventive maintenance action that sprays either atomized water or a mild abrasive solution containing coke into the engine to remove accumulated buildup. When routinely cleaned, engines have reduced fuel flow and a reduction in exhaust gas temperature, which may result in deferred maintenance. Engine washing is routinely done in the commercial sector for fuel

---

[1] Lockheed Martin, "The C-130J Super Hercules: Available Now, and Ready for Whatever the Future Holds," online, undated (b).

[2] Lockheed Martin, undated (b).

[3] FalconView, 2010.

conservation, and in 2010, the Air Force conducted a business case analysis on engine water washing and recommended implementation.[4] Currently, the Air Force washes all the MAF aircraft engines (e.g., C-5, C-17, C-130, KC-10, and KC-135), consequently, no additional fuel savings is available and, as a result, it was not included in our analysis.

Reserve Fuel Reduction

Commercial airlines have been taking advantage of a fuel planning technique referred to as re-dispatching.[5] By re-dispatching, airlines can reduce the amount of international contingency fuel required. This contingency fuel is specified as a percentage of flight time or planned fuel burn. Essentially, re-dispatching works by specifying a destination short of the final destination and calculating the reserves for each portion separately. However, the Air Force recently removed this requirement for C-17, formerly known as Category 1 Fuel. Currently, a Category 1 fuel requirement exists only for the C-130 and KC-135.[6] However, being a tactical airlifter, the percentage of time the C-130 spends over water is minimal. Since offloading fuel is the primary role of the KC-135, it would, for example, be impractical to re-dispatch on a CORONET mission when dragging fighters across oceans. Therefore, there is no significant opportunity for the Air Force to reduce reserve requirements by taking advantage of re-dispatching. Other fuel reserve reduction options were not considered here because of the associated flight safety issues.

Routing Improvement

We analyzed one day of FAA radar-track data covering all AMC flights in and around the national airspace. Excluding training missions, the data included 186 flights by C-5, C-17, and C-130 aircraft. We estimated average excess distance flown to be 34 nm (6 percent) and this was consistent across platforms. However, if hours for training and seasoning are a constraint, then the time saved by flying a direct route would have to be made up with additional sorties. This would result in an increased fuel use because more takeoffs were occurring even though flight hours remained constant.

---

[4] OC-ALC/FMC Acquisition Cost Division and OC-ALC/GKGAB F100/TF33/F119/F117 Program Management Section, "Air Force Jet Engine Water-Wash Business Case Analysis," August 26, 2010.

[5] Altus, 2009.

[6] AFI 11-2C-130, 2012; AFI 11-2C-130J, 2009; and AFI 11-2KC-135, 2008.

References

AFI—*See* Air Force Instruction.

Air Force Instruction 11-2C-5, *C-5 Operations Procedures*, Volume 3, §14.2.3, February 24, 2012.

———, 11-2C-17, Volume 3, *C-17 Operations Procedures*, §6.25.1, November 16, 2011. As of February 27, 2014:
http://www.bits.de/NRANEU/others/END-Archive/AFI11-2C-17V3(11).pdf

———, 11-2C-130, Volume 3, *C-130 Operations Procedures*, §14.2.5, April 23, 2012. As of February 27, 2014:
http://static.e-publishing.af.mil/production/1/aetc/publication/afi11-2c-130v3_aetcsup_i/afi11-2c-130v3_aetcsup_i.pdf

———, 11-2C-130J, Volume 3, *C-130J Operations Procedures*, §14.5.14, December 8, 2009.

———, 11-2KC-10, Volume 3, *KC-10 Operations Procedures*, §14.2.3, August 30, 2011. As of February 27, 2014:
http://www.altus.af.mil/shared/media/document/AFD-111108-034.pdf

———, 11-2KC-135, Volume 3, *C/KC-135 Operations Procedures*, §14.4.2.5, September 18, 2008, certified current October 15, 2010.

Air Mobility Command, *C-17 Cruise Speed Change,* Fuel Efficiency Office Bulletin 09-01, updated September 11, 2014.

———, no title, Fuel Efficiency Office Bulletin 09-02, updated September 11, 2014.

———, "Reduce APU Use," *Reduce APU Use Program Evaluation,* Scott AFB, Ill.: AMC/A4M, March 2012.

Air Transport Action Group, *Beginner's Guide to Aviation Efficiency,* November 2010. As of February 27, 2014:
http://www.enviro.aero/AviationEfficiency.aspx

Airbus, "The A320 New Engine Option," online, undated. As of February 27, 2014:
http://www.airbus.com/presscentre/hot-topics/a320neo/

———, "Getting to Grips with Fuel Economy," *Flight Operations Support & Line Assistance, Airbus Customer Services*, No. 3, July 2004.

———, "The A330/A340 Family Jetliners Benefit from Lower Maintenance Costs," press release, April 16, 2009. As of June 30, 2013:

http://www.airbus.com/presscentre/pressreleases/press-release-detail/detail/the-a330a340-
family-jetliners-benefit-from-lower-maintenance-costs/

Airlines for America (A4A), *2011 Economic Report*, undated. As of July 2, 2013:
http://www.airlines.org/Pages/A4A-Economic-Reports-of-the-U.S.-Airline-Industry.aspx

Allied Pilots Association, Submission of the Allied Pilots Association to the National
Transportation Safety Board Regarding the Accident of American Airlines Flight 587 at
Belle Harbor, New York, November 12, 2001, NTSB DCA02MA001, undated. As of August
14, 2013:
http://www.alliedpilots.org/Public/Topics/Issues/apa587finalsubmission.pdf

Altus, Steve, "Effective Flight Plans Can Help Airlines Economize," *Aero Quarterly* (Boeing),
Vol. 3, 2009, pp. 27–30. As of February 27, 2014:
http://www.boeing.com/commercial/aeromagazine/articles/qtr_03_09/article_08_1.html

AMC—*See* Air Mobility Command.

American Airlines Newsroom, "Fuel Smart," online, undated. As of March 3, 2014:
http://hub.aa.com/en/nr/media-kit/operations/fuelsmart

Anselmo, Joseph C., "Fuel Crisis Forces Airlines to Conserve, Drop by Drop," *Aviation Week &
Space Technology*, December 6, 2004, p. 54.

Aviation Partners Boeing, "Blended Winglets: Improved Takeoff Performance," online, undated
(a). As of February 27, 2014:
http://www.aviationpartnersboeing.com/winglets_itp.php

———, "Blended Winglets System List Prices," online, undated (b). As of February 27, 2014:
http://www.aviationpartnersboeing.com/products_list_prices.php

Ayan, Erdem, Hakan Telli, and Y. Volkan Pehlivanoglu, *Computational Investigation of C-130
Afterbody Drag Reduction by Finlets*, Istanbul, Turkey: Aeronautics and Space Technologies
Institute, Turkish Air Force Academy, undated.

Babikian, R., S. P. Lukachko, and I. A. Waitz, "The Historical Fuel Efficiency Characteristics of
Regional Aircraft from Technological, Operational and Cost Perspectives," *Journal of Air
Transport Management*, Vol. 8, No. 6, 2002, pp. 389–400.

Bargsten, Clayton J., and Malcolm T. Gibson, *NASA Innovation in Aeronautics: Select
Technologies That Have Shaped Modern Aviation*, Washington, D.C.: National Aeronautics
and Space Administration, NASA/TM-2011-216987, August 2011.

Bates, Matthew (Tech. Sgt.) "Every Drop Counts," Travis Air Force Base, Calif.: Defense Media
Activity, November 1, 2013. As of March 3, 2014:
http://www.amc.af.mil/news/story.asp?id=123369200

Bechert, D. W., M. Bruse, W. Hage, and R. Meyer, "Fluid Mechanics of Biological Surfaces and Their Technological Application," *Naturwissenschaften*, Vol. 87, No. 4, 2000, pp. 157–171.

Bechert, D. W., M. Bruse, W. Hage, J. G. T. Van der Hoeven, and G. Hoppe, "Experiments on Drag-Reducing Surfaces and Their Optimization with an Adjustable Geometry," *Journal of Fluid Mechanics*, Vol. 338, 1997, pp. 59–87. As of February 27, 2014: http://journals.cambridge.org/download.php?file=%2FFLM%2FFLM338% 2FS0022112096004673a.pdf&code=5dfd8ac0f5f7be7ad563057d88374391

Bednarz, Sean G., Anthony D. Rosello, Shane Tierney, David Cox, Steven C. Isley, Michael Kennedy, Chuck Stelzner, and Fred Timson, *Modernizing the Mobility Air Force for Tomorrow's Air Traffic Management System*, Santa Monica, Calif.: RAND Corporation, MG-1194-AF, 2012. As of February 27, 2014: http://www.rand.org/pubs/monographs/MG1194.html

Bender, Andrew, "American Airlines' Makeover: Design Pros Weigh In," *Forbes.com*, January 21, 2013. As of August 20, 2013: http://www.forbes.com/sites/andrewbender/2013/01/21/american-airlines-makeover-design-pros-weigh-in/

Bland, Maj Keith, "KC-135 Upgrades to Keep Aircraft Flying for Decades," Air Mobility Command Public Affairs, August 9, 2012. As of August 28, 2014: www.amc.af.mil/nws/story.asp?id=123313322

Boeing, "737 MAX," *New Air Plane*, online, undated (a). As of August 30, 2013: http://www.newairplane.com/737max/

———, "Airplane Characteristics for Airport Planning," online, undated (b). As of January 14, 2013: http://www.boeing.com/boeing/commercial/airports/plan_manuals.page

———, "C-17 Globemaster III Technical Specifications," online, undated (c). As of August 30, 2013: http://www.boeing.com/boeing/defense-space/military/c17/c17spec.page

———, "Defining 21st Century Flight, 787 Dreamliner," online, undated (d). As of August 30, 2013: http://www.newairplane.com/787/

———, "History: Globemaster III," online, undated (e). As of August 19, 2013: http://www.boeing.com/boeing/history/mdc/c-17.page

———, "The New 747-8," *newairplane.com*, online, undated (f). As of August 19, 2013: http://www.newairplane.com/747/design_highlights/#/technologically-advanced

Brueckner, J. K., and A. Zhang, "Airline Emission Charges: Effects on Airfares, Service Quality, and Aircraft Design," *Transportation Research Part B: Methodological*, Vol. 44, Issues 8–9, September–November 2010, pp. 960–971.

Butler, Amy, "KC-46A—First Photos . . .Sans Winglets," Ares, *aviationweek.com* blog, April 7, 2011. As of February 27, 2014:
http://www.aviationweek.com/blogs.aspx?plckblogid=blog:27ec4a53-dcc8-42d0-bd3a-01329aef79a7&plckpostid=blog:27ec4a53-dcc8-42d0-bd3a-01329aef79a7post:d59189c6-3deb-4549-8e0b-231b9253ae50

Carey, S., S. Chaudhuri, and T. Stynes, "Delta, US Airways Strike Positive Note," *Wall Street Journal*, October 24, 2012.

Carlton, Gary N, "Aircraft Corrosion Control: Assessment and Reduction of Chromate Exposures," Brooks Air Force Base, Tex.: Air Force Institute for Environment Safety, June 2000.

Cessna Aircraft Company, *Citation X: Specification & Description,* Units 750-0501 to TBD, Revision B, Preliminary, Wichita, Kan., March 2013. As of August 30, 2013:
http://www.cessna.com/~/media/Files/citation/x/xsd.ashx

Chow, Brian G., *The Peacetime Tempo of Air Mobility Operations: Meeting Demand and Maintaining Readiness*, Santa Monica, Calif.: RAND Corporation, MR-1506-AF, 2003. As of February 27, 2014:
http://www.rand.org/pubs/monograph_reports/MR1506.html

Code of Federal Regulations, Title 14—Aeronautics and Space, Section 119.3 Definitions, January 1, 2003.

Committee on Analysis of Air Force Engine Efficiency Improvement Options for Large Nonfighter Aircraft, National Research Council, *Improving the Efficiency of Engines for Large Nonfighter Aircraft,* Washington, D.C.: The National Academies Press, 2007.

Consulting Aviation Services, "VC Finlet: Reduces Drag on Upward Swept Fuselage Aircraft," online, undated. As of August 15, 2013:
http://www.casinc.us/innovations.php#Finlet

Crenshaw, Wayne, "Corrosion Office Helps Prolong Life of Aircraft," Robins Air Force Base, Ga.: 78th Air Base Wing Public Affairs, February 10, 2009. As of February 27, 2014:
http://www.afmc.af.mil/news/story.asp?id=123134611

Damodaran, Aswath, *Applied Corporate Finance*, 3rd ed., Danvers, Mass.: John Wiley & Sons, Inc., 2011.

DLA—*See* Defense Logistics Agency.

Defense Logistics Agency, "DLA Energy Standard Prices," online, undated (a). As of August 30, 2013:
http://www.energy.dla.mil/DLA_finance_energy/Pages/dlafp03.aspx

———, *Standard Fuel Prices in Dollars,* "FY 2013 President's Budget, FY 2013 Rates," online, undated (b). As of August 30, 2013:
http://www.energy.dla.mil/DLA_finance_energy/Documents/FY%202013%20Standard%20Prices%20(Effective%20Oct%201,%202012).pdf

Delta Airlines, "Environmental Fact Sheet," online, updated January 2010. As of March 3, 2014:
http://news.delta.com/index.php?s=18&item=83

Department of the Air Force, *United States Air Force Committee Staff Procurement Backup Book FY 2006/2007 Budget Estimates: Aircraft Procurement Air Force,* Vol. II, Washington, D.C.: SAF/FMB, February 2005.

Dion-Schwarz, Cynthia, Leon Hirsch, Phillip Koehn, Jenya Macheret, and Dave Sparrow, *FCS Vehicle Transportability, Survivability, and Reliability Analysis,* Alexandria, Va.: Institute for Defense Analysis, April 2005.

Disney, T. E., "C-5A Active Load Alleviation System," *Journal of Spacecraft and Rockets,* Vol. 14, No. 2, 1977, pp. 81–86.

DoE—*See* United States Department of Energy.

Drinnon, Roger, "'Vortex Surfing' Could Be Revolutionary," *Air Force Print News Today,* October 10, 2012. As of February 27, 2014:
http://www.afrc.af.mil/news/story_print.asp?id=123321615

Endres, Günter G., *The Illustrated Directory of Modern Commercial Aircraft,* Osceola, Wisc.: MBI Publishing Co., 2001.

Energy Information Administration, *Annual Energy Outlook 2014,* Appendix A – Table 3, undated. As of August 22, 2014:
http://www.eia.gov/forecasts/aeo/er/pdf/tbla3.pdf

Esler, David, "Question for Efficiency Driving Winglet Retrofits," *Aviation Week,* December 1, 2008.

FalconView, "Portable Flight Planning Software (PFPS)," online, circa October 2010. As of August 20, 2013:
http://www.falconview.org/trac/FalconView/wiki/PFPS

Farries, Pamela, and Chris Eyers, *Aviation CO2 Emissions Abatement Potential from Technology Innovation,* London: Committee on Climate Change, QINETIQ/CON/AP/CR0801111, October 14, 2008.

Federal Aviation Administration, "Atlantic Interoperability Initiative to Reduce Emissions (AIRE)," 2010 AIRE Workshop, Brussels, Belgium, 2010.

———, *Instrument Flying Handbook*, Oklahoma City, Okla.: United States Department of Transportation, 2012. As of September 28, 2013:
http://www.faa.gov/regulations_policies/handbooks_manuals/aviation/media/
FAA-H-8083-15B.pdf

Federation of American Scientists, "C-130 Hercules," online, updated February 20, 2000. As of August 19, 2013:
http://www.fas.org/man/dod-101/sys/ac/c-130.htm

FedEx Express, "FedEx and the Environment," fact sheet, online, June 2013. As of March 3, 2014:
http://news.van.fedex.com/sites/default/files/pressmaterials_file/June%202013%20FedEx%
20Express%20Environmental%20Fact%20Sheet.pdf

Garcia-Mayoral, Ricardo, and Javier Jimenez, "Drag Reduction by Riblets," *Philosophical Transactions of the Royal Society A*, Vol. 369, No. 1940, March 7, 2011, pp. 1412–1427.

GAO—*See* United States Government Accountability Office.

GRID-Arendal, "Aviation and the Global Atmosphere," online, 2001. As of March 4, 2014:
http://www.grida.no/publications/other/ipcc_sr/?src=/climate/ipcc/aviation/avf9-3.htm

Hansen, Dan, "Painting Versus Polishing of Airplane Exterior Surfaces," *AERO Magazine*, online, undated. As of August 20, 2013:
http://www.boeing.com/commercial/aeromagazine/aero_05/textonly/fo01txt.html

Hilkevitch, J., and J. Johnsson, "American Airlines' Effort to Cut Fuel Reserves Draws Fire from Pilots," *Los Angeles Times*, July 13, 2010.

Honeywell and Safran Aerospace Defence Security, "Electric Green Taxiing System," brochure, December 2011. As of March 4, 2014:
http://aerospace.honeywell.com/~/media/Brochures/etaxi-brochure.ashxfor additional information

ICAO—*See* International Civil Aviation Organization.

Insinna, Valerie, and Yasmin Tadjdeh, "Air Force Making Headway on Fuel Efficiency Goals," *National Defense Magazine*, June 2013.

Interagency Working Group on Social Cost of Carbon, United States Government, *Technical Support Document: Technical Update of the Social Cost of Carbon for Regulatory Impact Analysis Under Executive Order 12866*, May 2013. As of August 30, 2013:

http://www.whitehouse.gov/sites/default/files/omb/inforeg/social_cost_of_carbon_for_ria_2013_update.pdf

International Civil Aviation Organization, "ICAO DATA+ Glossary," online, undated. As of August 30, 2013:
http://www2.icao.int/en/G-CAD/Documents/GLOSSARY.pdf

———, *Continuous Descent Operations (CDO) Manual,* Doc. 9931, 1st ed., 2010 (a).

———, *ICAO Environmental Report 2010: Aviation and Climate Change*, 2010 (b). As of September 28, 3013:
http://www.icao.int/environmental-protection/Pages/EnvReport10.aspx

Jackson, P., ed., *Jane's All the World's Aircraft,* Jane's Information Group, 2012.

JetBlue, *2007 Environmental and Social Report*, online, undated (a). As of August 30, 2013:
http://www.jetblue.com/green/

———, *2012 Responsibility Report*, online, undated (b). As of August 22, 2013:
http://www.jetblue.com/green/

Jones, R. T., and T. A. Lasinski, "Effect of Winglets on the Induced Drag of Ideal Wing Shapes," Washington, D.C.: National Aeronautics and Space Administration, NASA Technical Memorandum 81230, September 1980.

Kennedy, Michael, Laura H. Baldwin, Michael Boito, Katherine M. Calef, James Chow, Joan Cornuet, Mel Eisman, Chris Fitzmartin, J. R. Gebman, Elham Ghashghai, Jeff Hagen, Thomas Hamilton, Gregory G. Hildebrandt, Yool Kim, Robert S. Leonard, Rosalind Lewis, Elvira N. Loredo, D. Norton, David T. Orletsky, Harold Scott Perdue, Raymond Pyles, Timothy L. Ramey, Charles Robert Roll, William L. Stanley, John Stillion, F. S. Timson, and John Tonkinson, *Analysis of Alternatives (AoA) for KC-135 Recapitalization: Executive Summary*, Santa Monica, Calif.: RAND Corporation, MG-495-AF, 2006. As of February 27, 2014:
http://www.rand.org/pubs/monographs/MG495.html

Kingsbury, Alex, "New Landings Save Airplane Fuel," *U.S. News and World Report*, July 2, 2008.

Koomey, Jonathan, *Comparative Analysis of Monetary Estimates of External Environmental Costs Associated with Combustion of Fossil Fuels*, Berkeley, Calif.: Lawrence Berkeley Laboratory, July 1990.

Larson, George, "How Things Work: Winglets," *Air & Space Magazine*, September 2001. As of March 4, 2014:
http://www.airspacemag.com/flight-today/how-things-work-winglets-2468375/?no-ist

Lee, Joosung Joseph, *Historical and Future Trends in Aircraft Performance, Cost, and Emissions*, dissertation, Cambridge, Mass.: Massachusetts Institute of Technology, 2000.

Lockheed Aeronautical System Company, "Meet the Hercules," *Service News*, Vol. 16, No. 1, January–March 1989.

Lockheed Martin, "C-5 Galaxy: Heavy Lifting," online, undated (a). As of August 19, 2013:
http://www.lockheedmartin.com/us/100years/stories/c5.html

———, "The C-130J Super Hercules: Available Now, and Ready for Whatever the Future Holds," online, undated (b). As of August 19, 2013:
http://www.lockheedmartin.com/us/news/trade-shows/singapore/sas-stories/sas-c-130.html

———, "C-130J Super Hercules Worldwide Fleet Soars Past 1 Million Flight Hours," Marietta, Ga., May 14, 2013. As of August 15, 2013:
http://www.lockheedmartin.com/us/news/press-releases/2013/may/0514aero-C-130j-million-flight-hours.html

Lubold, Gordon, "How Geese Will Save the Air Force Millions of Dollars," *Foreign Policy.com,* July 18, 2013. As of August 2, 2013:
http://killerapps.foreignpolicy.com/posts/2013/07/18/how_geese_will_save_the_air_force_millions_of_dollars

Lufthansa Technik, "Like a Shark in the Water: Innovative Lacquer System to Reduce Drag," online, October 2012. As of August 15, 2013:
http://www.lufthansa-technik.com/multifunctional-coating

MAC—*See* Military Airlift Command.

Marentic, F. J., and T. L. Morris, "1986 Drag Reduction Article," Minnesota Mining and Manufacturing Co. (assignee), U.S. Patent Number 5,133,516, St. Paul, Minn., July 28, 1992.

Mayer, Daryl, "First KC-46 Build Begins," Wright-Patterson Air Force Base, Ohio: 88th Air Base Wing Public Affairs, July 1, 2013. As of August 19, 2013:
http://www.afmc.af.mil/news/story.asp?id=123354525

McAndrews, Laura, "Fuel Efficiency Among Top Priorities in AMC's Energy Conservation," Scott Air Force Base, Ill.: Air Mobility Command Public Affairs, October 5, 2009. As of March 3, 2014:
http://www.amc.af.mil/news/story_print.asp?id=123171212

McCullough, Amy, "Energy Effectiveness," *Air Force Magazine*, January 19, 2011. As of August 16, 2013:
http://www.airforcemag.com/Features/airpower/Pages/box011911energy.aspx

McGowan, William A., "Aircraft Wake Turbulence Avoidance," paper presented at 12th Anglo-American Aeronautical Conference, Calgary, July 7–9, 1971. As of March 3, 2014: http://ntrs.nasa.gov/archive/nasa/casi.ntrs.nasa.gov/19720004302.pdf

McLean, J. Douglas, Dezso N. George-Falvy, and Peter P. Sullivan, "Flight-Test of Turbulent Skin Friction Reduction by Riblets," *Proceedings of International Conference on Turbulent Drag Reduction by Passive Means*, London: Royal Aeronautical Society, 1987, pp. 1-17.

Military Airlift Command, Navigation and Performance Division of Aircrew Standardization, "Birds Fly Free, MAC Doesn't," pamphlet, Scott AFB, Ill.: Navigation and Performance Division of Aircrew Standardization, February 10, 1976.

Moody, Elyse, "Focus on Fuel Savings," *Aviation Week & Space Technology*, March 1, 2012.

Morrison, J., P. Bonnefoy, R. J. Hansman, and S. Sgouridis, "Investigation of the Impacts of Effective Fuel Cost Increase on the U.S. Air Transportation Network and Fleet," *10th AIAA Aviation Technology, Integration, and Operations (ATIO) Conference*, Fort Worth, Tex., September 2010.

Mouton, Christopher A., James D. Powers, Daniel M. Romano, Christopher Guo, Sean Bednarz, and Caolionn O'Connell, *Fuel Reduction for the Mobility Air Forces: Executive Summary*, Santa Monica, Calif.: RAND Corporation, RR-757/1-AF, 2015. As of February 2015: http://www.rand.org/pubs/research_reports/RR757z1.html

Mouton, Christopher A., and Stephen L. Graham, "Investigation of Drag Reduction in Turbulent Flow over Riblet Covered Surfaces," paper prepared for AIAA Student Conference, Region IV, Houston, TX, 2000.

Mouton, Christopher A., David T. Orletsky, Michael Kennedy, and Fred Timson, *Reducing Long-Term Costs While Preserving a Robust Strategic Airlift Fleet: Options for the Current Fleet and Next-Generation Aircraft, Appendix A*, Santa Monica, Calif.: RAND Corporation, MG-1238-AF, 2013. As of February 27, 2014: http://www.rand.org/pubs/monographs/MG1238.html

Nakahara, Alex, Tom G. Reynolds, Thomas White, Chris Maccarone, and Ron Dunsky, "Analysis of a Surface Congestion Management Technique at New York JFK Airport," paper prepared for 11th AIAA Aviation Technology, Integration and Operations (ATIO) Conference, Virginia Beach, Va., September 2011.

National Research Council, Committee on Assessment of Aircraft Winglets for Large Aircraft Fuel Efficiency, *Assessment of Wingtip Modifications to Increase the Fuel Efficiency of Air Force Aircraft*, Washington, D.C.: The National Academies Press, 2007.

NASA—*See* National Aeronautics and Space Administration.

National Advisory Committee for Aeronautics, *The Spanwise Distribution of Lift for Minimum Induced Drag of Wings Having a Given Lift and a Given Bending Moment,* Moffett Field, Calif.: Ames Aeronautical Laboratory, Technical Note 2249, December 1950.

National Aeronautics and Space Administration, "Winglets Save Billions of Dollars in Fuel Costs," online, undated. As of August 16, 2013:
http://spinoff.nasa.gov/Spinoff2010/t_5.html

———, "Sky Surfing for Fuel Economy," July 24, 2003. As of August 14, 2013:
http://www.nasa.gov/missions/research/vortex.html

Naval Air Systems Command, *Cleaning and Corrosion Control,* Volume I: *Corrosion Program and Corrosion Theory,* Patuxent River, Md., NAVAIR 01-1A-509-1, March 1, 2005. As of February 27, 2014:
http://www.robins.af.mil/shared/media/document/AFD-091006-032.pdf

Nikoleris, Tasos, Gautam Gupta, and Matthew Kistler, "Detailed Estimation of Fuel Consumption and Emissions During Aircraft Taxi Operations at Dallas/Fort Worth International Airport," *Transportation Research Part D: Transport and Environment,* Vol. 16, No. 4, June 2011, pp. 302–308.

Norton, Daniel M., Donald Stevens, Yool Kim, Scott Hardiman, Somi Seong, Fred Timson, John Tonkinson, Duncan Long, Nidhi Kalra, Paul Dreyer, Artur Usanov, Kay Sullivan Faith, Benjamin F. Mundell, and Katherine M. Calef, *An Assessment of the Addition of Winglets to the Air Force Tanker Fleets,* Santa Monica, Calif.: RAND Corporation, MG-895-1-AF, January 2012, not available to the general public.

OC-ALC/FMC Acquisition Cost Division and OC-ALC/GKGAB F100/TF33/F119/F117 Program Management Section, "Air Force Jet Engine Water-Wash Business Case Analysis," August 26, 2010.

OMB—*See* United States Office of Management and Budget.

Pahle, Joe, Dave Berger, Michael W. Venti, James J. Faber, Chris Duggan, and Kyle Cardinal, "A Preliminary Flight Investigation of Formation Flight for Drag Reduction on the C-17 Aircraft," briefing, Washington, D.C.: National Aeronautics and Space Administration, NASA 20120007201, March 7, 2012. As of March 3, 2014:
http://ntrs.nasa.gov/archive/nasa/casi.ntrs.nasa.gov/20120007201.pdf

Pappalardo, Joe, "Vortex Surfing: Formation Flying Could Save the Air Force Millions on Fuel," *Popular Mechanics,* July 17, 2013. As of August 14, 2013:
http://www.popularmechanics.com/technology/military/planes-uavs/vortex-surfing-formation-flying-could-save-the-air-force-millions-on-fuel-15703217

Parra, Art, fuel efficiency manager for FedEx, telephone communication with the author, March 26, 2013.

PASSUR Aerospace, "Arrival Management," fact sheet, online, undated. As of February 3, 2014:
http://www.passur.com/passur-arrival-management.htm

———, "PASSUR Aerospace Reports Revenue Increase of 22% for Fiscal Year 2010," Stamford, Conn., January 31, 2011. As of February 3, 2014:
http://www.passur.com/55f365a2-dfc7-48dc-803a-eded743e4a19/news-and-events-press-release-detail.htm

Peet, Yulia, and Pierre Sagaut, "Turbulent Drag Reduction Using Sinusoidal Riblets with Triangular Cross-Section," American Institute of Aeronautics and Astronautics Paper AIAA-2008-3745, prepared for 38th AIAA Fluid Dynamics Conference and Exhibit, Seattle, Washington, June 23–26, 2008.

Peeters, P. M., J. Middel, and A. Hoolhorst, *Fuel Efficiency of Commercial Aircraft: An Overview of Historical and Future Trends*, Amsterdam, Netherlands: Peeters Advies/National Aerospace Laboratory, NLR-R-2005-669, November 2005.

Prandtl, L., "Über Tragflügel des Kleinsten Induzierten Widerstandes," *Zeitschrift für Flugtechnik und Motorluftschiffahrt*, Vol. 24, 1933, pp. 305–306.

Pratt & Whitney, *Operating Instructions for the PW4000 Series Commercial Turbofan Engines*, online, undated. As of June 2008:
http://www.pw.utc.com/PW400094_Engine

Ray, Elizabeth L., "Air Traffic Control," FAA Order JO7110.65U, Washington, D.C.: Federal Aviation Administration, December 16, 2011.

Ray, Ronald J., Brent R. Cobleigh, M. Jake Vachon, and Clinton St. John, "Flight Test Techniques Used to Evaluate Performance Benefits During Formation Flight," Edwards AFB, Calif.: National Aerospace Laboratory, Dryden Flight Research Center, NASA/TP-2002-210730, August 2012.

Raymer, Daniel P., *Aircraft Design: A Conceptual Approach*, 3rd ed., Reston, Va.: American Institute of Aeronautics and Astronautics, Inc., 1999.

———, *Aircraft Design: A Conceptual Approach*, 4th edition, Reston, Va.: American Institute of Aeronautics and Astronautics, 2006.

Reed, Dan, "Can Fuel Hedges Keep Southwest in the Money?" *USA TODAY,* online, updated July 24, 2008. As of March 3, 2014:
http://usatoday30.usatoday.com/money/industries/travel/2008-07-23-southwest-jet-fuel_N.htm

Reid, Laurence, "Will the North American Commercial Jet Fleet Experience Growth in the Next Ten Years or Will the Next Decade Be One Solely of High Replacement and Minimal Growth for This Region?" *Ascend World Wide*, April 12, 2013. As of August 19, 2013: http://www.ascendworldwide.com/2013/04/will-the-north-american-commercial-jet-fleet-experience-growth-in-the-next-ten-years-or-will-the-nex.html

Rhodes, Jeff, "Tweak My Ride," *Code One* (Lockheed Martin), March 7, 2012. As of August 15 2013: http://www.codeonemagazine.com/article.html?item_id=91

Roberson, B., and S. S. Pilot, "Fuel Conservation Strategies: Cost Index Explained," *Aero Quarterly* (Boeing), Vol. 2, 2007, pp. 26–28.

Rosello, Anthony D., Sean Bednarz, Michael Kennedy, Chuck Stelzner, F. S. Timson, and David T. Orletsky, *Assessing the Cost-Effectiveness of Modernizing the KC-10 to Meet Global Air Traffic Management Mandates*, Santa Monica, Calif.: RAND Corporation, MG-901-AF, 2009. As of February 27, 2014: http://www.rand.org/pubs/monographs/MG901.html

Rosello, Anthony D., Sean Bednarz, David T. Orletsky, Michael Kennedy, Fred Timson, Chuck Stelzner, and Katherine M. Calef, *Upgrading the Extender: Which Options Are Cost-Effective for Modernizing the KC-10?* Santa Monica, Calif.: RAND Corporation, TR-901-AF, 2011. As of February 27, 2014: http://www.rand.org/pubs/technical_reports/TR901.html

Roth, John P., Office of the Under Secretary of Defense (Comptroller), "FY 2013 Department of Defense (DoD) Military Personnel Composite Standard Pay and Reimbursement Rates," memorandum, April 9, 2012.

Rowbotham, Jim, "Coatings for Composites," *AERO Magazine*, February 13, 2012. As of August 20, 2013: http://www.aero-mag.com/features/41/20122/1270/

Schonaeuer, Scott, "'Winglets' Could Save Air Force Millions on Fuel," *Stars and Stripes*, October 1, 2007. As of March 4, 2014: http://www.stripes.com/news/winglets-could-save-air-force-millions-on-fuel-1.69425

Serbu, Jared, "Air Force Meets Fuel Efficiency Goal Several Years Early," *federalnewsradio.com*, March 22, 2013. As of August 19, 2013: http://www.federalnewsradio.com/395/3259612/Air-Force-meets-fuel-efficiency-goal-several-years-early

Sherwin-Williams Aerospace Coatings, "Military Aerospace Coatings," product data, online, undated. As of September 28, 2013: http://www.swaerospace.com/products/exterior/topcoats/military/

Shinkman, Paul D., "'Vortex Surfing' Could Save Military Millions," *U.S. News and World Report*, August 13, 2013. As of August 14, 2013:
http://www.usnews.com/news/articles/2013/08/13/vortex-surfing-could-save-military-millions

Smith, Brian, Patrick Yagle, and John Hooker, "Reduction of Aft Fuselage Drag on the C-130 Using Microvanes," 51st AIAA Aerospace Sciences Meeting Including the New Horizons Forum and Aerospace Exposition, Grapevine, Texas, January 7–10, 2013. As of August 30, 2013:
http://arc.aiaa.org/doi/abs/10.2514/6.2013-105

Smith, Kyle, "C-130 Hercules, Fuel Efficiency Initiatives," Lockheed Martin Advanced Development Programs, undated (a).

———, "Fuel Efficiency Initiatives," briefing, Lockheed Martin Advanced Development Programs, undated (b). As of August 30, 2013:
http://www.lockheedmartin.com/content/dam/lockheed/data/aero/documents/global-sustainment/product-support/2012HOC-Presentations/Wednesday/Wed%201600%20Fuel%20Efficiency%20Initiatives-Kyle%20Smith.pdf

Smoot, Harold, "AWBS: Automated Weight & Balance System," Lockheed Martin Corporation, undated.

Sohn, Myong Hwan, and Jo Won Chang, "Visualization and PIV Study of Wing-Tip Vortices for Three Different Tip Configurations," *Aerospace Science and Technology*, Vol. 16, No. 1, January–February 2012, pp. 40–46.

Stibbe, Matthew, "U.S. Air Force Will Save $50M with iPad Electronic Flight Bags," *Forbes.com*, May 30, 2013. As of February 3, 2014:
http://www.forbes.com/sites/matthewstibbe/2013/05/30/u-s-air-force-will-save-50m-with-ipad-electronic-flight-bags/

Sturkol, Scott T., "AMC Fuel Efficiency Office Shows How 'Efficiency Promotes Effectiveness,'" Scott Air Force Base, Ill.: Air Mobility Command Public Affairs, January 5, 2011. As of August 19, 2013:
http://www.af.mil/DesktopModules/ArticleCS/Print.aspx?PortalId=1&ModuleId=850&Article=114505

Sullivan, Michael, Bruce Fairbaim, Keith Hudson, John Krump, May Jo Lewnard, Don Springman, Roxanna Sun, and Robert Swierczek, *KC-46 Tanker Aircraft: Acquisition Plans Have Good Features but Contain Schedule Risk*, Washington, D.C.: Government Accountability Office, GAO-12-366, 2012.

Szodruch, J., "Viscous Drag Reduction on Transport Aircraft," AIAA Paper 91-0685, 29th Aerospace Sciences Meeting, Reno, Nevada, January 7–10, 1991.

Technical Order 1-1B-50, *Basic Technical Order for USAF Aircraft: Weight and Balance,* Tinker AFB, Okla.: 557 ACSS/GFEAC, April 1, 2008. As of March 3, 2014: http://www.tinker.af.mil/shared/media/document/AFD-061214-031.pdf

Timson, Fred, *Analysis of Alternatives (AoA) for KC-135 Recapitalization: Appendix E— Acquisition Costs for New-Design Alternatives,* Santa Monica, Calif.: RAND Corporation, MG-460-AF, December 2005, not available to the general public.

United Airlines, "United Eco-Skies: A Commitment to the Environment," online, December 11, 2011. As of March 4, 2014: https://hub.united.com/en-us/Videos/Company/Pages/eco-skies.aspx

———, "United Airlines is First to Install Split Scimitar Winglets," press release, July 17, 2013. As of March 4, 2014: http://newsroom.unitedcontinentalholdings.com/2013-07-17-United-Airlines-is-First-to-Install-Split-Scimitar-Winglets

United Kingdom Civil Aviation Authority, "Basic Principles of the Continuous Descent Approach (CDA) for the Non-Aviation Community," London: Environmental Research and Consultancy Department of the Civil Aviation Authority, undated.

United Parcel Service, "Fuel Management and Conservation at the UPS Airlines," online, undated. As of August 22, 2013: http://www.pressroom.ups.com/Fact+Sheets/Fuel+Management+and+Conservation+at+the+UPS+Airlines

United States Coast Guard, Office of Aviation Forces (CG-711), "HC-130H: Hercules," online, last modified on June 28, 2013. As of August 15, 2013: http://www.uscg.mil/hq/cg7/cg711/c130h.asp

United States Department of Defense, *Air Force Aircraft Painting and Corrosion Control,* Washington, D.C., Inspector General Audit Report No. 96-062, January 1996.

———, Executive Services Directorate, "Weight and Balance Clearance Form F—Transport," August 1996. As of September 11, 2014: http://www.dtic.mil/whs/directives/infomgt/forms/eforms/dd0365-4.pdf

———, *Fiscal Year (FY) 2013 President's Budget: Flying Hour Program (PA),* an extract of the Programmed Data System (PDS), Washington, D.C., February 2012, not available to the general public.

United States Department of Energy, *Air Force Achieves Fuel Efficiency Through Industry Best Practices,* Washington, D.C., DOE/GO-102012-3725, December 2012. As of September 22,

2013:
http://www1.eere.energy.gov/femp/pdfs/af_fuelefficiency.pdf

United States Department of Transportation Federal Aviation Administration Flight Standards Service, *Aircraft Weight and Balance Handbook*, Washington, D.C., 2007.

United States Energy Information Administration, "U.S. Gulf Coast Kerosene-Type Jet Fuel Spot Price FOB," online, release date August 28, 2013. As of September 22, 2013: http://www.eia.gov/dnav/pet/hist/LeafHandler.ashx?n=pet&s=eer_epjk_pf4_rgc_dpg&f=d

United States General Accounting Office, *C-5A Wing Modification: A Case Study Illustrating Problems in the Defense Weapons Acquisition Process*, Washington, D.C., March 1982.

United States Government Accountability Office, *Aviation and Climate Change*, Washington, D.C., GAO-09-554, 2009. As of September 28, 2013: www.gao.gov/new.items/d09554.pdf

OMB—*See* United States Office of Management and Budget.

United States Office of Management and Budget, *Guidelines and Discounted Rates for Benefit-Cost Analysis of Federal Programs*, OMB Circular No. A-94, Appendix C, "Table of Past Years Discount Rates," Washington, D.C., revised December 2012. As of September 28, 2013:
http://www.whitehouse.gov/omb/circulars_a094/a94_appx-c

Vedantham, A., et al., "Aircraft Emissions: Current Inventories and Future Scenarios," in *Aviation and the Global Atmosphere: A Special Report of IPCC Working Groups I and III*, Cambridge, Mass.: Cambridge University Press, January 1999.

Vortex Control Technologies, "Finlets Technology—Applicable Aircraft: C130 /L100 Hercules," online, undated. As of February 27, 2014:
http://www.vortexct.com/products/finlets/aircraft/c130l-100-hercules/

———, *2013 Program Price List*, January 2013. As of February 27, 2014:
http://www.vortexct.com/wp-content/uploads/2013/07/VCT-Program-Pricing-January-2013.pdf

Walsh, Michael J., "Riblets as a Viscous Drag Reduction Technique," *AIAA Journal*, Vol. 21, No. 4, 1983, pp. 485–486.

Warsop, Clyde, "Current Status and Prospects for Turbulent Flow Control, Aerodynamic Drag Reduction Technologies," in Peter Thiede, ed., *Proceedings of the CEAS/DragNet European Drag Reduction Conference,* Potsdam, Germany, June 19–21, 2000.

Warwick, Graham, "Lockheed Developing Winglets for C-130, C-5," *Aerospace Daily & Defense Report*, October 6, 2011, p. 3.

———, "C-17s Go Surfing, to Save Fuel," *aviationweek.com* blog post, October 12, 2012. As of August 2, 2013:
http://www.aviationweek.com/Blogs.aspx?plckPostId=Blog:27ec4a53-dcc8-42d0-bd3a-01329aef79a7Post:63a5d3dc-0182-415f-ad8a-35dbbed3da2b

———, "NASA, Boeing Study Flexible Wing Control," *Aviation Week & Space Technology*, January 23, 2013 (a).

———, "How Many Bin Bags to Empty an A340?" *Aviation Week*, blog post, March 20, 2013 (b). As of August 16, 2013:
http://www.aviationweek.com/Blogs.aspx?plckPostId=Blog:7a78f54e-b3dd-4fa6-ae6e-dff2ffd7bdbbPost:596045d7-2aef-4b10-b9b5-bd78c93b351f

Whitcomb, Richard T., *A Design Approach and Selected Wind-Tunnel Results at High Subsonic Speeds for Wing-Tip Mounted Winglets,* Hampton, Va.: Langley Research Center, NASA Technical Note D-8280, July 1976. As of February 27, 2014:
http://ntrs.nasa.gov/archive/nasa/casi.ntrs.nasa.gov/19760019075.pdf

Wilkinson, S. P., et al., "Turbulent Drag Reduction Research at NASA Langley: Progress and Plans," *International Journal of Heat and Fluid Flow*, Vol. 9, No. 3, September 1988.

Wilson, Ian, and Florian Hafner, "Benefit Assessment of Using Continuous Descent Approaches at Atlanta," paper presented at 24th Digital Avionics Systems Conference, Washington, D.C., October 30, 2005.